BUILDING CONSTRUCTION

Design Aspects of Leakage and Seepage Free Buildings

BUILDING CONSTRUCTION

Design Aspects of Leakage and Seepage Free Buildings

Krishna Kant
M.Tech, F.I.E., F.I.V., MICA, MIRC, MIBC
Former UN Expert and Chief Engineer, Central PWD

Tata McGraw Hill Education Private Limited
NEW DELHI

McGraw-Hill Offices

New Delhi New York St Louis San Francisco Auckland Bogotá Caracas
Kuala Lumpur Lisbon London Madrid Mexico City Milan Montreal
San Juan Santiago Singapore Sydney Tokyo Toronto

Tata McGraw-Hill

Published by Tata McGraw Hill Education Private Limited,
7 West Patel Nagar, New Delhi 110 008

ISBN-13: 978-1-25-906121-9
ISBN-10: 1-25-906121-3

Vice President and Managing Director: *Ajay Shukla*
Publishing Manager—Professional: *Praveen Tiwari*
Editorial Researcher—Science, Technology and Computing: *Sushil Gupta*
Asst. General Manager—Sales and Business Development—Professional: *S Girish*
Deputy Marketing Manager—Science, Technology and Computing: *Rekha Dhyani*
General Manager—Production: *Rajender P Ghansela*
Manager—Production: *Reji Kumar*

Typeset at SP Graphics, 130 DDA Flat, 1st Floor, Todapur (Opp. Pusa Institute), New Delhi-110 012 and printed at Satnat Printers, 312 EPIP, HSIDC, Kundli, Sonepat, Haryana.

Cover Printer: Sanat Printers

RXCLCRYHRYBZA

The McGraw·Hill Companies

To
My Parents

FOREWORD

Though lot of literature and detailed specifications/standards are available for various civil engineering works, dealing exhaustively with aspects like planning, structural designs, actual execution; the attention paid to the finishing items has been lacking. One of the areas which need immediate attention is treatment of wet areas of house such as kitchens, toilets bathrooms, etc. Occurrence of leakage/seepage, as soon as the buildings/installations are put to use; also is not an uncommon sight. Such construction works, not only increase maintenance cost of building, but also invite adverse criticism from the users and public as well. It is with this issue in view that Er. Krishna Kant has taken up on himself the onerous task of consolidating various problems/requirements related to wet areas at one place and highlighting factors that should be given due attention for their improvement.

The present book is written with the goal to achieve "Buildings free from seepage/dampness" and increase the awareness among architects, engineers, contractors, builders and students. Accordingly efforts have been made to link the various contents like planning in the early stages from layout of various wet areas in building and their structural requirements, basic objectives in selection of materials, points to be taken care of during execution and their subsequent maintenance, etc. The requirements/suggestions are lucidly explained with the help of sketches. Care has also been taken to adopt the standards applicable to our country.

I am sure that this book will be of much help for effective execution of the work and bridge the existing gap.

V.R. Vaish

V.R. Vaish

New Delhi

Director General of Work (Retd.)
Central Public Works Department

चन्द्रशेखर प्रसाद
महानिदेशक

C.S. Prasad
Director General

सत्यमेव जयते

केन्द्रीय लोक निर्माण विभाग
निर्माण भवन, नई दिल्ली-110011
Central Public Works Department
Nirman Bhawan, New Delhi-110011
Tel.: 23062556/1317 Fax: 23061884
E-mail: cpwd_dgw@nic.in

दिनांक / Dated: 19th March 2012

OVERVIEW

All of us, whether practicing professionals or users, are faced with the common problem of leakages and seepages in Buildings. A good amount of literature is available on various aspects of Civil Engineering and Building Construction but the emphasis needed to this problem has somehow been lacking. Er. Krishna Kant has had firsthand experience of dealing with these problems having been actively involved with maintenance and upkeep of large number of residential and office buildings in different parts of India and abroad during his long meritorious tenure with CPWD. With this in view he had published a book Titled "towards Better Living-Trouble Free Kitchens and Toilets." This book was very well received and was found to be useful by practicing engineers.

The present book is an updation of the earlier book and now encompasses broader subject of all buildings and steps that will help in delivery of leakages and seepages free buildings, save users from ugly damp patches, scarred looks and unhealthy surroundings. Various materials and methodologies available for waterproofing have been discussed with suggestions as to the best solutions. Emphasis is on practical approach to the problems. The norms/standards that need to be followed for various installations, fittings and fixtures within the building have been explained with detailed drawings/sketches.

All working details and suggested methodologies have been lucidly explained with very clear and well drawn sketches to make their application clear and easy. I am sure details provided, solutions suggested and lucid presentation of the subject matter of the book will motivate and benefit all practicing Engineers/Architects/Builders and users.

I congratulate Er. Krishna Kant on coming up with a comprehensive book on this very important aspect of day to day life which has been seeking attention all along. I do hope that all professionals will make best use of this publication.

C.S. PRASAD
Director General of Works
C.P.W.D.
Nirman Bhawan, New Delhi

PREFACE

Building construction is art as well as science. Construction of buildings involves attention to various aspects such as space planning, aesthetic looks, structural safety, functionality etc. As buildings are to be used extensively by us, they have necessarily to be serviceable and are required to be user friendly. For making the building functional, services of the buildings have an important role to play. It is, however, unfortunate that the emphasis which is required to be paid to planning, designing and execution of services has not been always available. It is more true in the case of planning, detailing and installation of water supply, sanitary fittings/ fixtures and drainage.

Whereas lot of literature is available on different aspects of building designing and construction; lack of literature on making buildings leakage and Seepage free is glaring. Keeping this vacuum in mind a book Titled "Trouble Free Kitchens & Toilets" was written and published by the author in 2000. The book was well received and suggestions kept pouring in for enhancement of scope of the book from mere Kitchen and Toilets to buildings as a whole. The present book on "Leakages Seepages Free Buildings" is an endeavour in this direction and proposes to fill in the existing gap.

Leakage and seepages in any building not only give bad appearance but also deteriorate the basic structure reducing its useful life of building and cause unhealthy environment which may lead to diseases. To ensure trouble free wet areas and stop ingress of water into the building as well as avoid unwanted dampness, steps are required to be taken at different stages of building planning, execution and use.

The book analyzes causes that lead to ingress of water from outside the building or from within the buildings and then proceeds on to suggest measures that need to be taken to avoid the same and also dwells on material selection, their proper use and correct methodologies to achieve the objective. Common problems met with by the Engineers/Architects, builders and users have been discussed in detail with suggested precautions and solutions.

The subject matter has been illustrated with detailed sketches for better clarity and understanding. Standard details for fittings and fixtures have also been incorporated to help execution of work and comfort to users. Methods of waterproofing for basements, Toilets/Kitchens and Terraces etc normally required in buildings have been dealt in detail.

Emphasis in the book is on practical aspects that help the planners and executing persons. Check lists have been added to facilitate supervision of important aspects of work. It is hoped that the present book will be found useful by practicing Architects/Engineers/Builders and home owners as well as students pursuing building construction related aspects.

I thank all those who have helped in preparation of the book and also various authors/agencies whose publications have been read and acknowledged. Suggestions for improvement and making the book still more useful are welcome.

KRISHNA KANT

CONTENTS

CHAPTER 1

CAUSES AND SOURCES OF LEAKAGES AND SEEPAGES

1.1 INTRODUCTION

Shelter has been and continues to be one of the three basic needs of mankind. We have come a long way since Stone Age of building construction. We are now able to construct sky scrapers and intelligent buildings. With passage of time our needs have undergone a sea change and so has the building construction. Gone are the days when kitchens were kept far away from house to save it from smoke/gases. Similarly no one now thinks of locating bathing and toilet facilities far away from the residential area as with modern techniques we are easily able to take care of the foul smells and disposal of wastes.

While deciding on any residential or commercial accommodation, one not only looks for good location, aesthetic looks but also reliability and functionality of the units. Leakage and seepages in any building not only give bad appearance but also deteriorate the basic structure reducing its useful life and cause unhealthy environment which may lead to diseases. To ensure trouble free wet areas precautionary measures are required to be taken in the early stages of building planning and execution itself and followed by proper upkeep and maintenance.

With the introduction of water carriage system for disposal of wastes, the dry system has been phased out. The facilities which we need most and at odd hours, need no longer be away from the areas of living. On account of this planning and execution services for the comfort of users without any negative aspects has become more important.

However, when there are advantages of a system, there are some precautions or do's and don'ts to follow for getting the maximum advantage of the system with least of problems. The present day water borne system has made our lives much easier and comfortable but gross misuse of it and lack of proper application of mind or attention to detailing has resulted in unhygienic conditions and unsightly appearances. In a good number of buildings one could point out the location of kitchens or toilets/baths even without going into the buildings for convenience sake these areas are being referred to wet areas as water is extensively used in these areas. When proper attention is not paid at planning stage, at execution stage problems of leakage/seepage is being built in. These are further aggravated by lack of awareness or attention required for proper maintenance and upkeep of these areas. 70% of the maintenance problems are caused due to leakages and seepages in buildings. Since this segment of building construction has been neglected for too long and it has been resulting in not only unsightly buildings but also in discomfort of living, dampness, unhygienic conditions and at times to diseases, in this book it is proposed to deal with these areas, their specific requirements, factors needing attention for proper design, construction and upkeep.

Ingress of water can be from below (subsoil water), from above and walls, i.e., rain water and/or from inside the building owing to improper water supply, sewerage and drainage. Attention is required to be paid to their prevention at every stage of planning, execution and maintenance to have comfortable, functionally efficient and aesthetic buildings.

Some of the jobs which are of repetitive nature need standardisation. Typical standardised details for such installations and areas giving precise details for their location for best human comfort, ease of accessibility and utility and have been prepared and are depicted in fully dimensioned figures in the book. All dimensions are in mm unless specified. These, when adopted, will ensure better construction, ease of use and will result in hygienic environments.

1.2 SOURCES OF LEAKAGES AND SEEPAGES

Normal sources of leakages and seepages in buildings are:

- Through Soil and Masonry by Capillary action.

- Water pressure exerted on structural members and their inability to withstand the same.
- External sources such as rain/storm.
- Internal water sources such as leaking pipes/improper fittings.
- From roof on account of inadequate waterproofing/slopes of roof or improper drainage.
- Flooding of areas and improper or inadequate drainage.

Leakages and seepages in wet areas (Fig. 1.1 and 1.2) are normally caused due to:

1. The joint between flushing cistern and flushing pipe.
2. Junction of the flushing pipe with the WC pan.
3. Junction of WC and traps with the branch pipe.

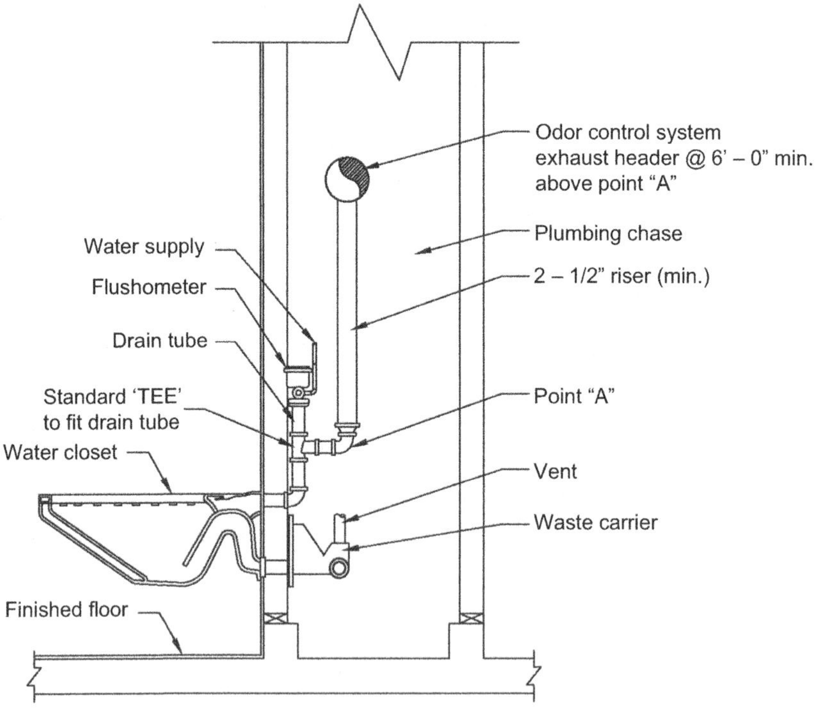

Fig. 1.1 *Sources of leakages in water closets and flushing cisterns.*

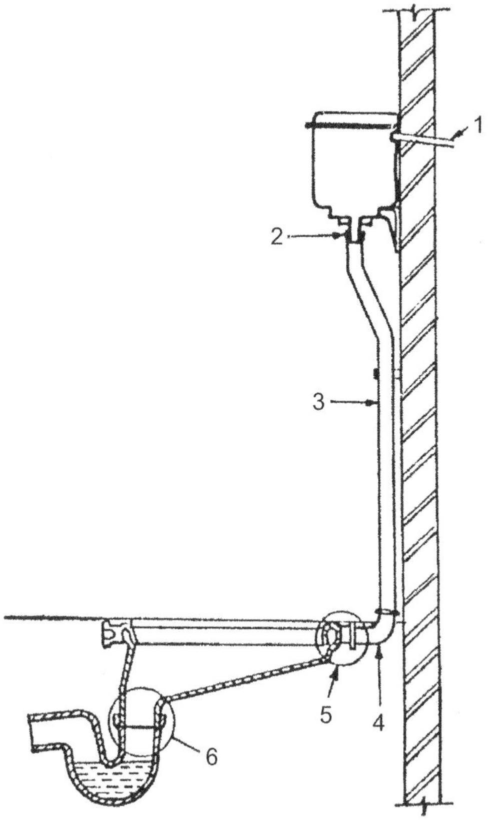

Notes:

1. Incorrect placement of overflow pipe
2. Joint between flushing cistern and flushing pipe
3. Downtake flushing pipe, when embedded
4. The bend of the flushing pipe which is embedded on floor
5. Junction of flushing pipe with the WC pan
6. Trap junction

Fig. 1.2 *Sources of leakages in water closets and flushing cisterns*

4. Depressed reinforced concrete slab to accommodate the pan and trap, if not done with proper quality concrete; inadequate waterproofing and testing.
5. Joints between flooring tiles.

6. Incorrect placement of overflow pipe.
7. Floor trap junction or improperly fixed floor trap.
8. Faulty pipe joints.
9. Cut out in the structure for the branch pipe.
10. Floor and walls where shower splashes water.
11. Junction of door frame and floor on account of eventual rotting of door frame.
12. Concealed piping and fittings: joints and faulty pipes.
13. Improper slopes in floors and sanitary pipes and lack of adequate drainage pipes.
14. Area surrounding the kitchen sink.
15. Junction joint between kitchen platform and the wall.
16. Junction of chajja and wall.
17. Throwing of solid wastes in WC's.
18. Improper and inadequate opening for draining off water.
19. At the junction of the water tank wall and the various pipe fittings in the tank walls.
20. Under the water tank on account of usage of scour valve.
21. Overflow from the water tanks.
22. From cracks in facades, terraces, roofs and sunken floors.
23. Use of porous finishes and brick work on exterior walls.
24. Lack of periodic checks and cleaning arrangements.

1.3 CAUSES OF LEAKAGES AND SEEPAGES

Before any action can be taken to attend and rectify the problem of leakage/seepage of water in buildings; its causes need to be ascertained first. These could be due to:

Sl. No.	*Structural Causes*	*Non-Structural Causes*
1	Cracks on the top surface of slab	Slope not proper
2	Cracks on the bottom of slab due to excess deflection of slab	Inadequate rain water pipes improper drainage
3	Cracks at column supports	Parapet wall not in good condition
4	Excess porosity in the slab, bad quality concrete poor quality brickwork.	Internal cracks, honeycombs (finished with ordinary mortar), porosity, etc., which cannot be seen by eye
5	Expansion joint in the slab	Mechanical installations like cooling towers, AC units, etc., on the terrace
6	Inadequate design of the slab	Aggressive climatic conditions
7	Spalling of concrete due to corrosion in the slab	Thermal extremes
8	Cracks due to differential settlement of a building	Damp proof course not provided or not effective

To take care of these problem areas, precautions are required to be taken in planning and selection of materials both during their use and execution of work. If proper detailing is done before execution, materials of good quality only are selected and precautions during various stages of work, as detailed in this book are taken, our constructions are likely to be trouble free as far as seepage/leakage or dampness is concerned.

CHAPTER 2

PLANNING STAGE

Prevention is better than cure cannot be truer than in case of leakage and seepages.

To minimise the problems of dampness, leakage and seepage, best course is to take precautions in the initial stages of the project. These include:

1. Steps required at the Planning Stage
2. Steps required to be taken during the Execution Stage.
3. Care during utilisation of utilities.

2.1 THE NEED

A beautifully designed building with very good specifications and properly landscaped can be an ugly sight due to leakage/seepages in wet areas. Its avoidance should start at planning and design stage itself.

Just as it is essential to plan and design various living areas of a building, it is also important that location, layout and designing of wet areas, i.e., kitchen, bath and W.C. their disposal shafts should be given due attention while planning various activities and installations.

Maximum advantage of efforts put in for prevention of leakage/ seepage is achieved from the measures adopted in the Planning and Design Stage of the Project.

2.2 LOCATION OF WET AREAS

India being a tropical country the byelaws of most local bodies makes it mandatory that wet areas be located on exterior of building, permitting natural ventilation and ease for disposal of wastes. It is good practice to keep these service areas such that at least one wall is facing the open area. Architects can plan the location keeping principles of Vastu Shastra in mind but without compromising with the above principle of ensuring natural light, view and adequate ventilation. In hotels and fully air-conditioned premises there is a growing tendency to have the toilets on either side of central corridors. Proper mechanical ventilation, adequate size of shafts with access to them must be ensured in such cases.

2.3 ARCHITECHTURAL DETAILING

(a) It is essential that the attention is paid to likely sources of ingress of water and provision made for the same.

(b) Protection to openings such as windows/doors should be provided.

(c) Impervious material used on exterior to prevent water seepage through walls/claddings.

(d) Good quality windows with proper sealants used to take care of rain water.

(e) Similarly chhajjas should have adequate slope and flow of water should not be obstructed. Joint of chhajja with wall should be specially detailed and treated. Drip course can be replaced with the detail shown for better water disposal. (See Fig. 3.7).

(f) Sills of windows should slope outwards.

(g) Terrace copings should be treated with impervious material and must slope on to the terrace.

(h) Expansion joints must be carefully detailed as these, more often than not, are big source of water ingress. Joints at different locations need specific treatment and detailing and have been shown in Figs 3.8 and 3.9.

(i) Box type projections that can retain water should be avoided.

2.4 LAYOUTS

As soon as the drawing is received the layout for kitchen, bath and toilets should be carefully worked out along with details of the disposal pipes

(See Fig. 2.1). Levels need to be worked out. Efforts should be that each fixture disposes off directly in the disposal pipes in the shaft. Waste water pipes and installations are installed:

- Within depressed slabs or
- Below the structural slab when these are hidden from direct view by providing false ceiling.

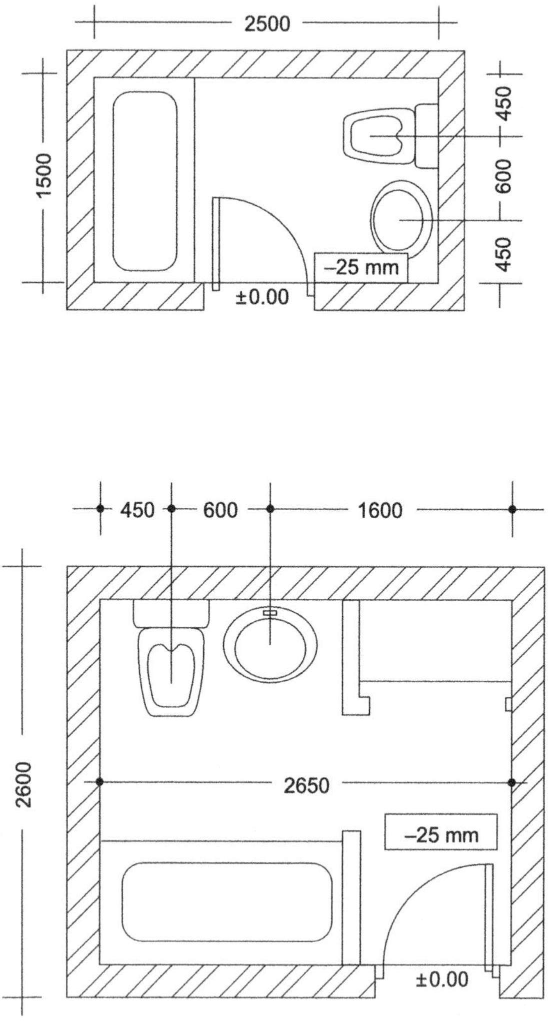

Fig. 2.1 *Typical toilet layout*

The older practice of accommodating waste pipes fittings and fixtures within depressions made availability of these installations to subsequent repairs very difficult. Even sources of leakage and seepages could not be identified easily. These drawbacks have led to laying of waste pipes below the floor which can be attended to from lower floor and actual points of leakages and seepages can be visually located.

Depth of the depression of the structural slabs, where so designed; should be worked out after the layout and levels are finalised. Many a times adequate space is not available for accommodating the floor traps, required slopes of long pipes in depressed floors and unwanted compromise is required to be made.

2.5 KITCHEN LAYOUTS

Kitchen is the most used area in any home and house lady spends most of her active hours in the area. Hence design and location of kitchen are most important to the house. Attention is required to be paid to proper location of washing area, cooking range and refrigerator. A few typical layouts of kitchens are given in Figs 2.2–2.5 which illustrates kitchen configurations. Layout has to meet the various functional, structural, spatial and other design considerations. Special attention is required to be paid to:

- The type, size and location of fixtures and appliances.
- The relationship and distance between the kitchen work centres.
- The amount of work surface and storage space required.
- Plumbing requirements and their location.
- Electrical lighting scheme and appliances to be used.
- Heating sources.
- *Vaastu* considerations.

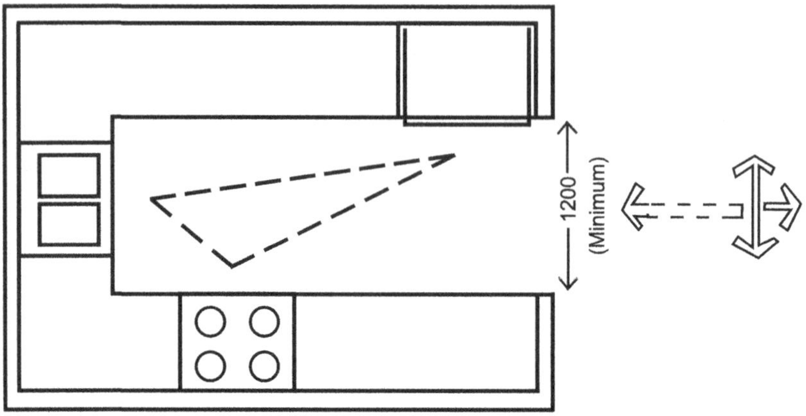

U Plan

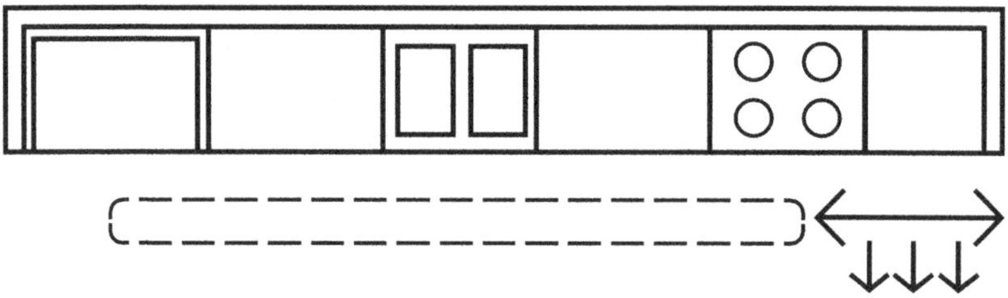

Single wall plan

Fig. 2.2 *Kitchen details: Typical layouts*

Kitchens should be planned and equipped to include the following: (See Figs 2.5 and 2.6).

(a) A sink and a draining board.

(b) The drainer should not overlap a return.

(c) A cooker energy outlet.

(d) A cooker space which should not be located beneath a window.

(e) A clear space of at least 100 mm between the cooker space and any return.

(f) The sink and cooker space should be separated by at least 600 mm at a return.

(g) Clear spacing of at least 1.0 m should be provided in front of fittings.

(h) Working platform should not be less than 600 mm wide. It should project by 20 mm beyond slabs and should have minimum joints. Minor slope away from the wall should be provided.

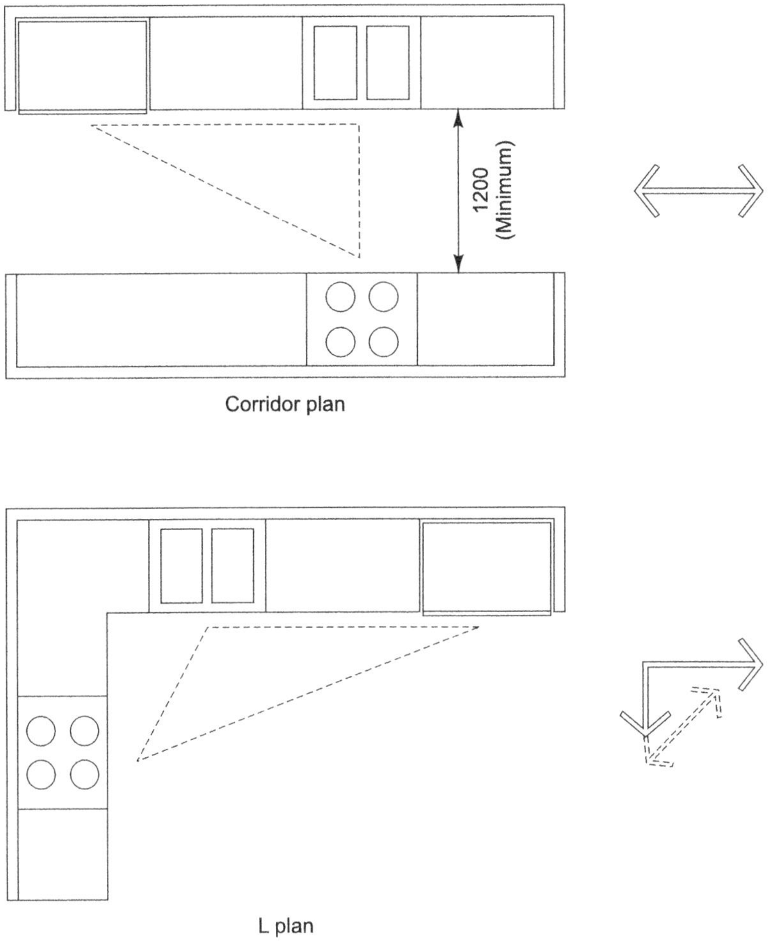

Fig. 2.3 *Kitchen details: Typical layouts*

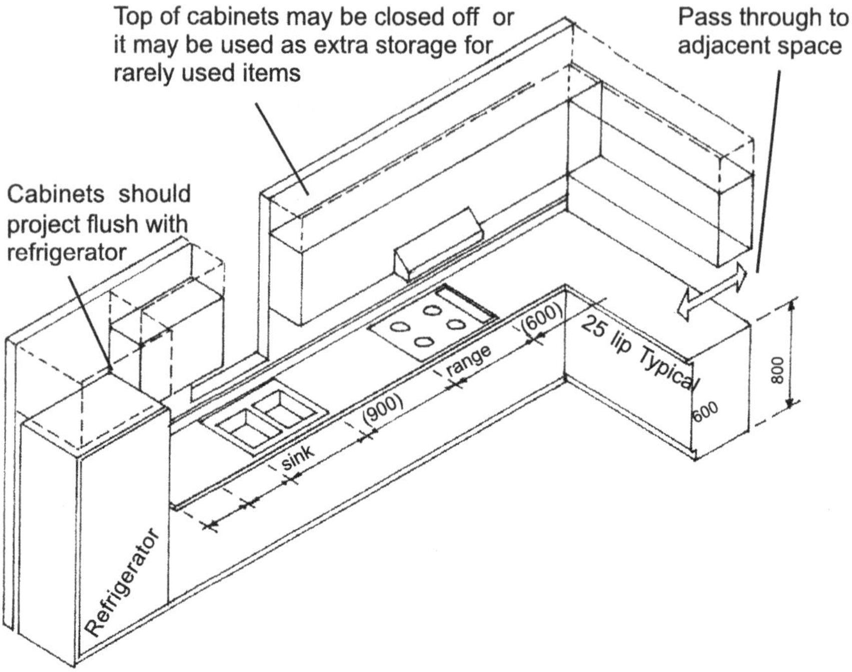

Fig. 2.4 *Kitchen details*

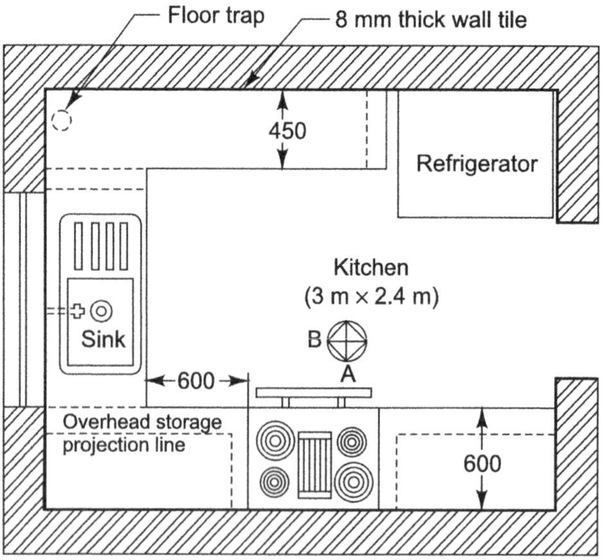

Fig. 2.5 *Kitchen plan*

(i) Height of working platform should be 750–800 mm from finished floor level.

(j) The bottom platform should be 80 mm high and recessed by 80 mm from the finished face of working platform (See Fig. 2.7) so that it does not obstruct the feet of working person.

(k) Where there is no piped gas supply, space for gas cylinder should be provided. Bottom platform in that location may have to be avoided.

(l) Proper arrangements must be made for exhaust fan or hood should be provided for extraction of smoke.

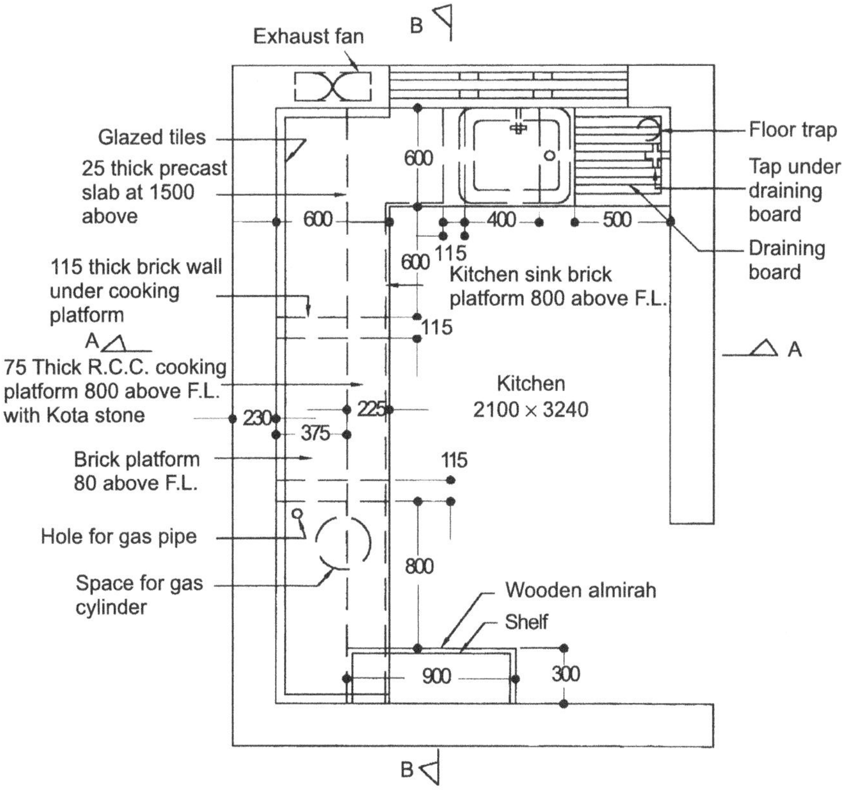

Fig. 2.6 *Typical kitchen layout*

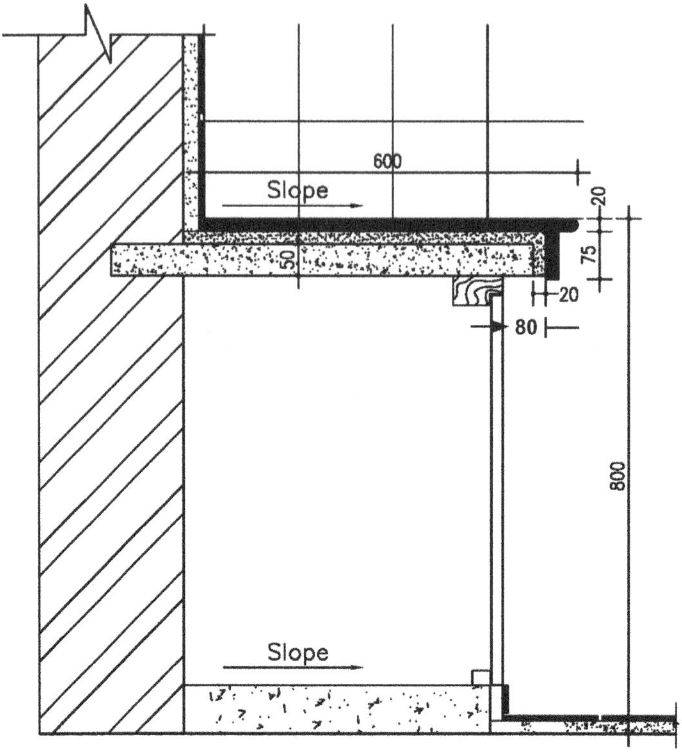

Fig. 2.7 *Kitchen platform*

(m) The total work surface space should not be less than 1.5 m long including drainer. No work space to be less than 300 mm.

(n) Adequate light/power points for electric gadgets to be provided. (See Figs 2.8 and 2.9). Gas outlet should be away from electric outlets.

(o) Adequate storage space for kitchen items and their retrieval needs special attention. Keeping in view our conditions typical alternatives of shelves to be provided in kitchen are given in Figs 2.9 and 2.10. These spacing will facilitate working of the user.

(p) When designing and providing Modular kitchens, which are quite popular now, due attention must be paid to choice of material. It must be waterproof and fire resistant.

(q) From safety point of view, it is a good practice to locate light switch for kitchen outside the kitchen.

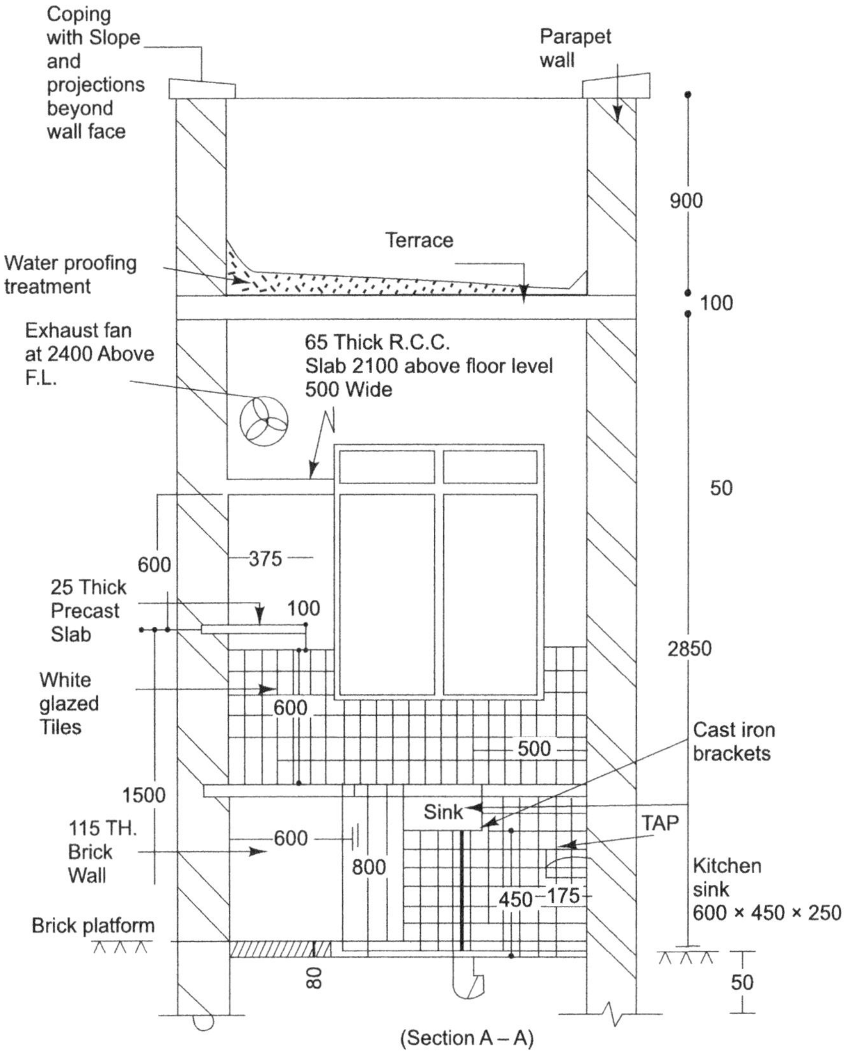

Fig. 2.8 *Kitchen details*

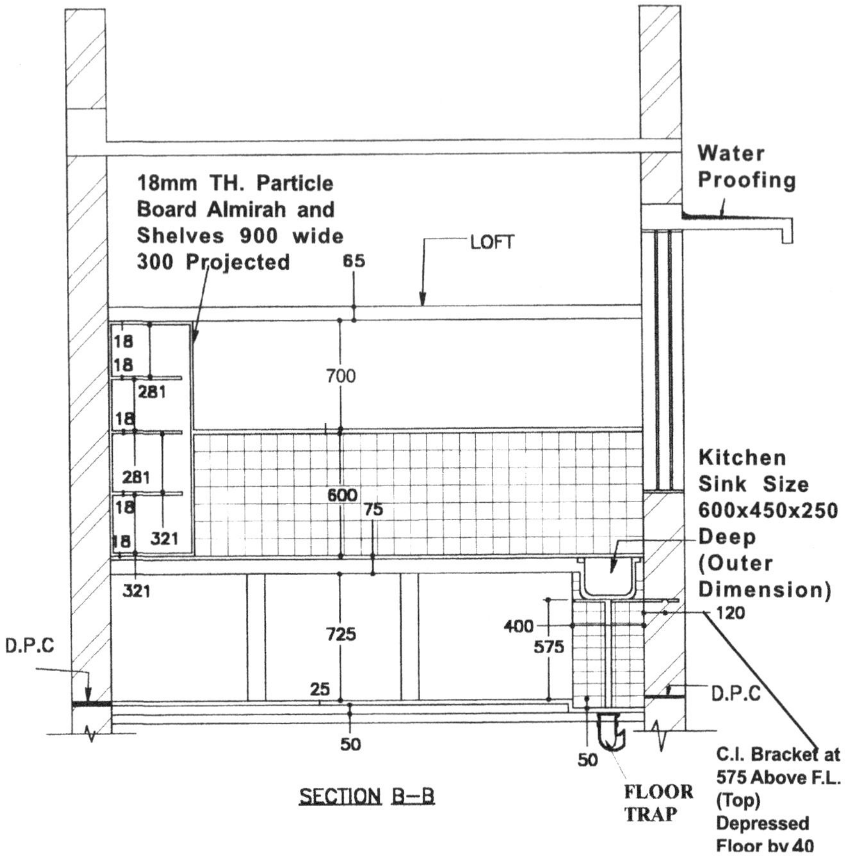

Fig. 2.9 *Kitchen details*

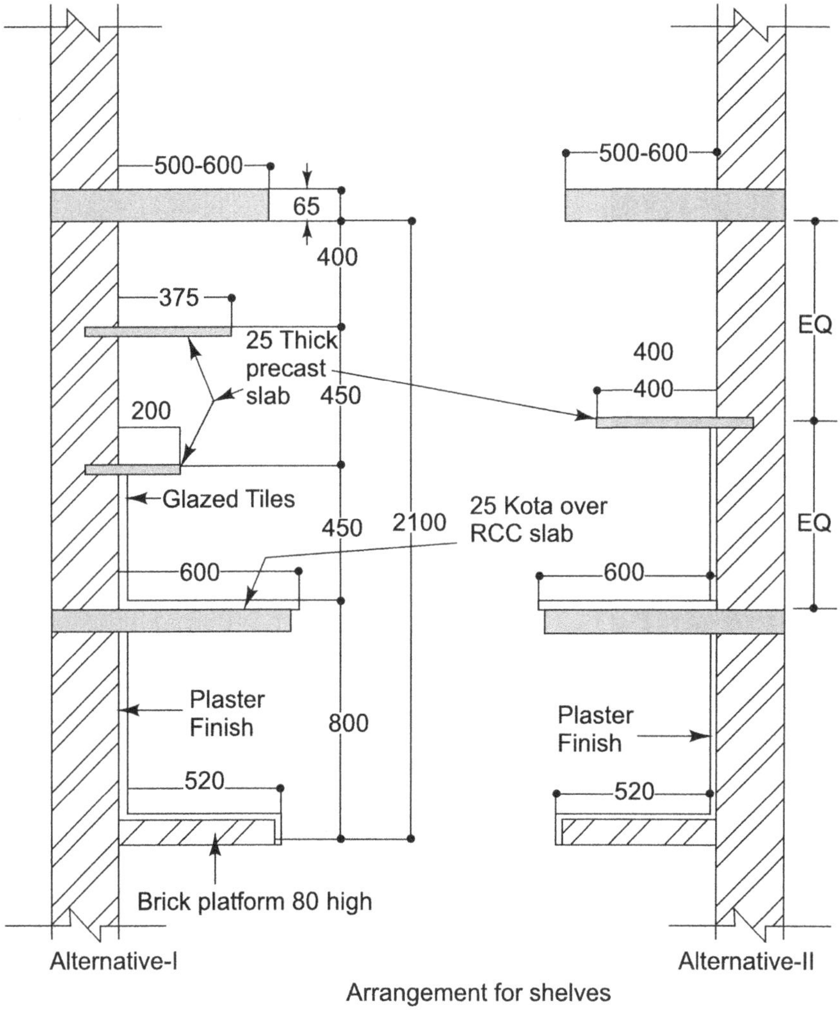

Fig. 2.10 *Kitchen details*

2.5.1 Vaastu considerations for kitchens

It is desirable to locate kitchen in the south eastern portion of the house. Door can be located preferably in the North, east or north east opening clockwise. While cooking the face should be towards east and hence cooker to be placed accordingly. Cooking Gas may be located in south east of kitchen but not in front of main door of the kitchen and sink be installed in north east. Exhaust fan can be on eastern side wall. Water filter can be installed in north east and dish washer in north west. Refrigerator may be located in south east corner.

Main kitchen platform should be in east and south east corner of the main building or flat. If there is a dining table in the kitchen it should be in the north west or the west side.

2.6 TOILETS LAYOUTS

While planning the toilets special consideration should be given to vision, sound and odour considerations. Vision could be obstructed by the

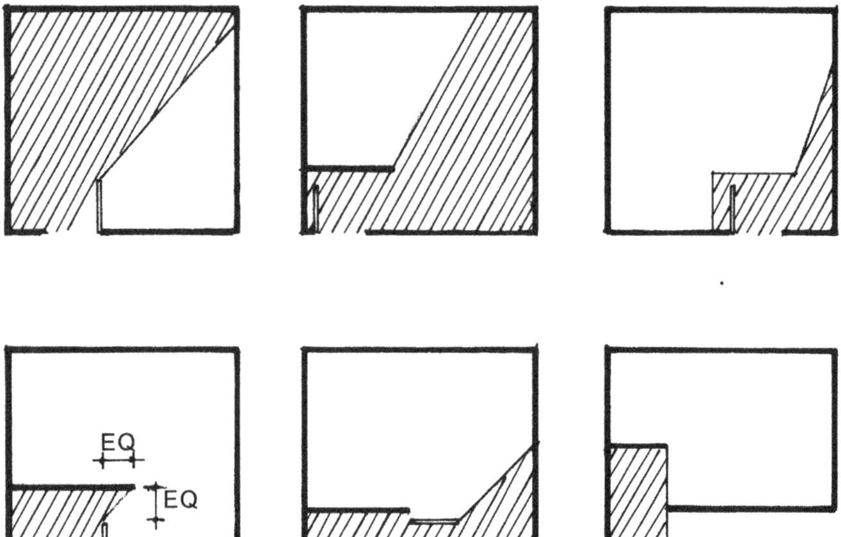

Various screening arrangements for small installations showing the area visible from outside

Fig. 2.11 *Toilet details*

configuration of entrance. (See Fig. 2.11) Doors could open inwards and be hung so that the appliances and users are screened as far as possible.

The doors of adjacent male and female wash rooms should not be next to each other as this is psychologically disturbing and aggravates vision problems. Consideration should also be given to positioning of mirrors and to gaps created by the hinges. Doors should be self closing where ever possible.

Special attention should be paid at planning stage to ensure noise reduction and also ventilation to minimise odour. Manually switched fans which run for a set period after being switched off are useful in domestic situations.

Considerable attention is required to be paid to the provisions of WC and washing facilities for the disabled, especially those using wheeled chairs as it is difficult to negotiate the wheel chairs. Proper hand rails and supports are required to be provided with due care and their users in mind. General requirements are shown in Figs 2.12, 2.13 and 2.14.

2.6.1 Vaastu Considerations for Toilets

Toilets should be located in the Northwest corner of the building or in the Northwest corner of the room. Where it is not possible then Southeast location can be used for toilets. Central south and central west are other options if first two are not feasible. Toilets in Northeast and in North and East should be avoided. Mirrors in the toilets should be square or rectangular and may be on North or East walls.

Toilet seats to be from south to north or west to east and may be located in west, south or north west side of west. Doors can be on east or north east. Slope of the flooring of toilets and outlets should be in the east or the north. Geyser can be installed in south east of the bath room. Bath tub, where provided, can be on west side with head side in south. A washing machine can be placed in the North West or South East portion of the bathroom.

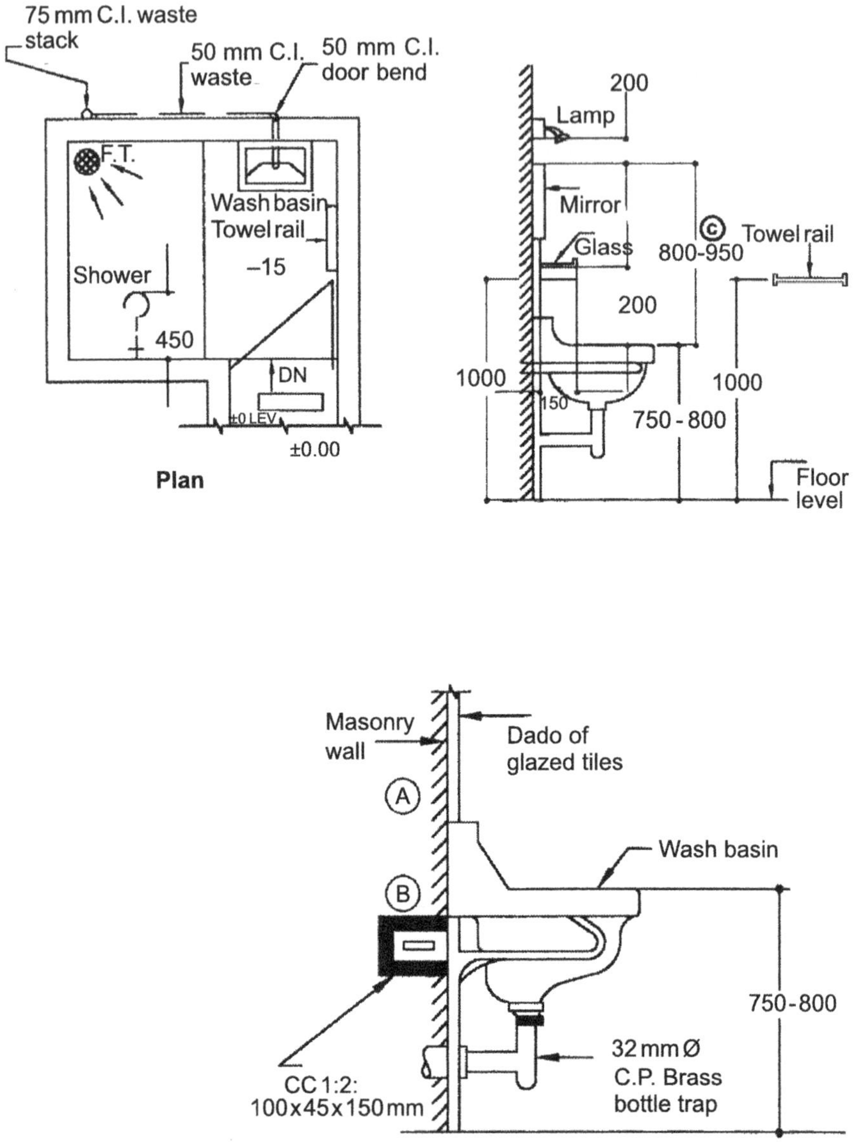

Fig. 2.12 *Typical layout for toilets*

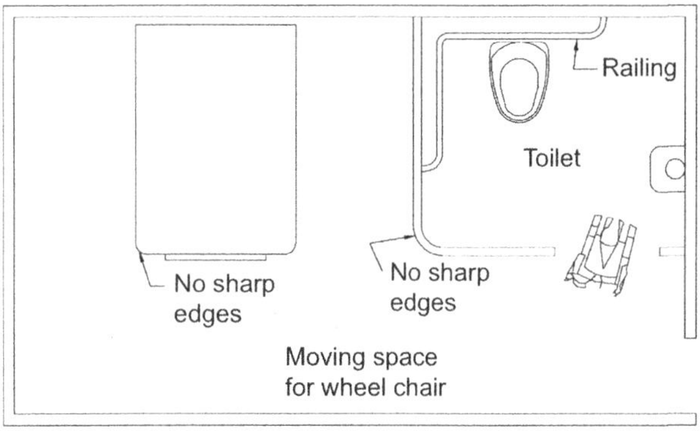

Fig. 2.13 *Toilet nearest to bedroom*

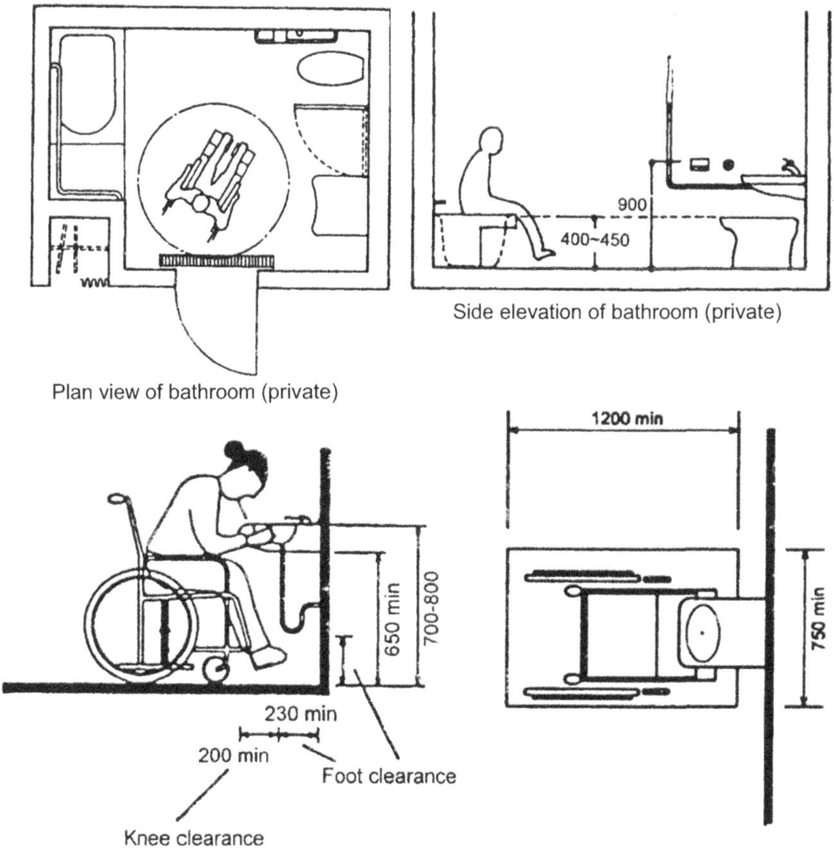

Fig. 2.14 *Bathroom for disabled*

2.7 SHAFTS

Disposal of wastes from wet areas is eventually taken through shafts. Shafts, which house the service pipes, continue to be the most neglected part of any building. This is one of the major reasons that the walls of shafts are a constant source of water seepage or dampness. Needless to say that greater attention is be paid to these service areas as well.

One need not be allergic to the looks of services. Efforts should rather be to make them look neat, clean and presentable. However, when shafts are provided, it must be ensured that, they are of adequate size, keeping in view the number of pipes, their location in shaft. Most of the time, at designing stages the fact, that these services will have to be maintained for the life of building, is forgotten. We always try to economise in shaft areas. At times spaces are so small that even construction pipes cannot be laid. One may learn a thing or two from the liberal sizes of shafts provided in our old buildings like Rashtrapati Bhawan, Parliament House, etc., where great attention has been given to the requirements of maintenance personnel who will go about attending their jobs during course of occupancy of the buildings.

Proper access for maintenance personnel, working platforms at levels which make it easy to stand and work on the joints of the pipes should be ensured. Metal rungs are not of much use in multi-storeyed buildings. It is desirable to provide access to shafts from each floor for ease of working. Working platform should not be at the floor level as the joints of pipes which need attention are normally just below floor level. Platform should be provided keeping location of joints and working space in mind. A person should be able to stand and attend such joints.

Shafts accommodate water supply, sewer, sullage, storm water drainage pipes and other services. Number of pipes will depend upon the fittings and fixtures served on the particular floor and floors above and are required to be kept in mind while deciding on location and size of shaft. Shafts are smaller than 1200 x 900 mm may not be used. Wet and dry shafts should run separately.

2.7.1 Shaft designs

Following additional points need to be kept in view while designing shafts:-

(a) Size and location of shafts to be decided keeping in view:

  (i) The number of pipes they are likely to cater

  (ii) Space required for maintenance personnel and work area required

  (iii) Approach to shaft and services therein

  (iv) Economy of service pipes.

(b) Working platforms and access:

  (i) Proper access and working space for maintenance personnel.

  (ii) Working platform should be provided keeping location of joints pipes that may need maintenance.

(c) Shaft finishes:

  (i) Mostly shafts remain open and are susceptible to ingress of water hence walls of shafts should be plastered with richer mortar of mortar mix 1:4 (1 cement: 4 coarse sand) using waterproofing compound. Plastering should be done prior to fixing of pipes as later plastering work cannot be done properly. When pipes are fixed, all damaged portions should be properly repaired to make them water proof.

  (ii) Proper floor at bottom of shaft, with adequate slope must invariably be given to drain off any water finding its way in.

  (iii) Shafts should have proper light and ventilation.

  (iv) It is good practice to provide raised cover on shaft area at terrace level, with fibre glass sheets, to prevent direct rain but allowing light and ventilation.

  (v) The shaft should be painted with water proof cement paint.

CHAPTER 3

DESIGN STAGE

3.1 DESIGN REQUIREMENTS

Good sanitary network has to be designed and installed to prevent foul air into the building, minimise frequency of blockage, adequate pipe access for attending to repairs, provide efficient discharge from appliances and minimise chance of flooding. Sanitary discharge pipe work should be kept as short as possible.

If proper slopes are ensured in the floors and the pipes, problems of water seepage/leakage get minimised. This can be achieved by doing proper planning of service with due care to see that:

(i) Floor of wet area vis-a-vis adjoining area is depressed by not less than 25 mm. (See Fig. 3.1)

(ii) Provision for waterproofing of the structural slab is made.

(iii) Adequate depth for accommodating floor traps, pipes, slope is provided. Normally 300 mm depression for kitchen/bath and European type W.C. installations and 600 mm for Indian Type W.C's are adequate. For public toilets and bigger areas exact depths have to be worked out based on the layout and distances from shaft, etc.

(iv) When services are taken below structural slab, the pipes above structural slab and below structural slab should have adequate slopes. Sleeves at pre decided locations should be provided and openings in the floors/walls of wet areas should be sealed by methods described in Fig. 3.2.

(iv) Effort should be to take out waste by shortest and straight route.

(v) Number of joints under the floors to be avoided or kept to minimum.

(vi) The structural slab must be cast with rich mix concrete, properly compacted and should be provided with slope towards shaft.

(vii) Typical details to be followed and dimensions to be adopted are given in (Figs 3.2–3.6).

(viii) While planning plumbing works, separate controls for each segment should be provided so that all services are not disrupted while carrying out repairs to one segment/fixture.

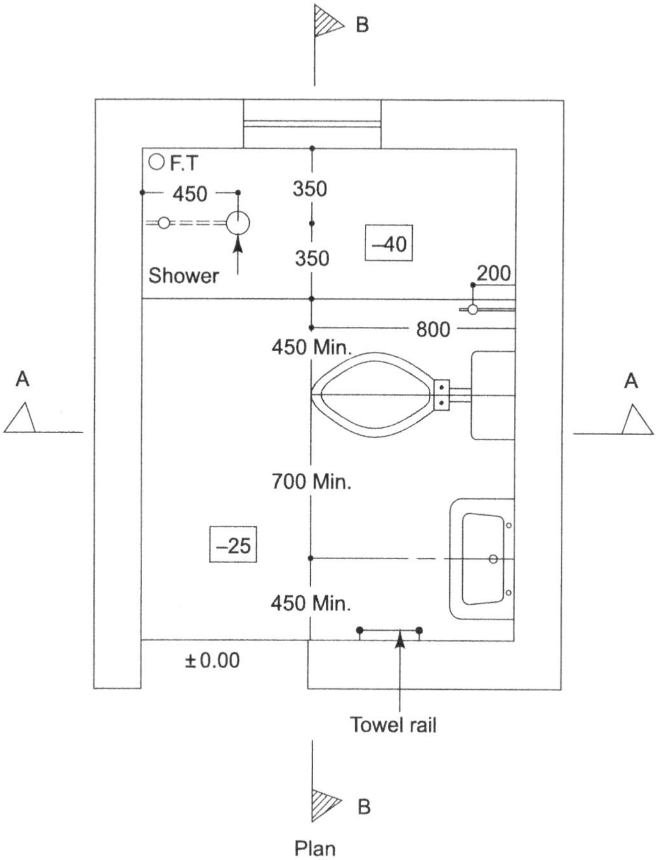

Fig. 3.1 *Toilet details*

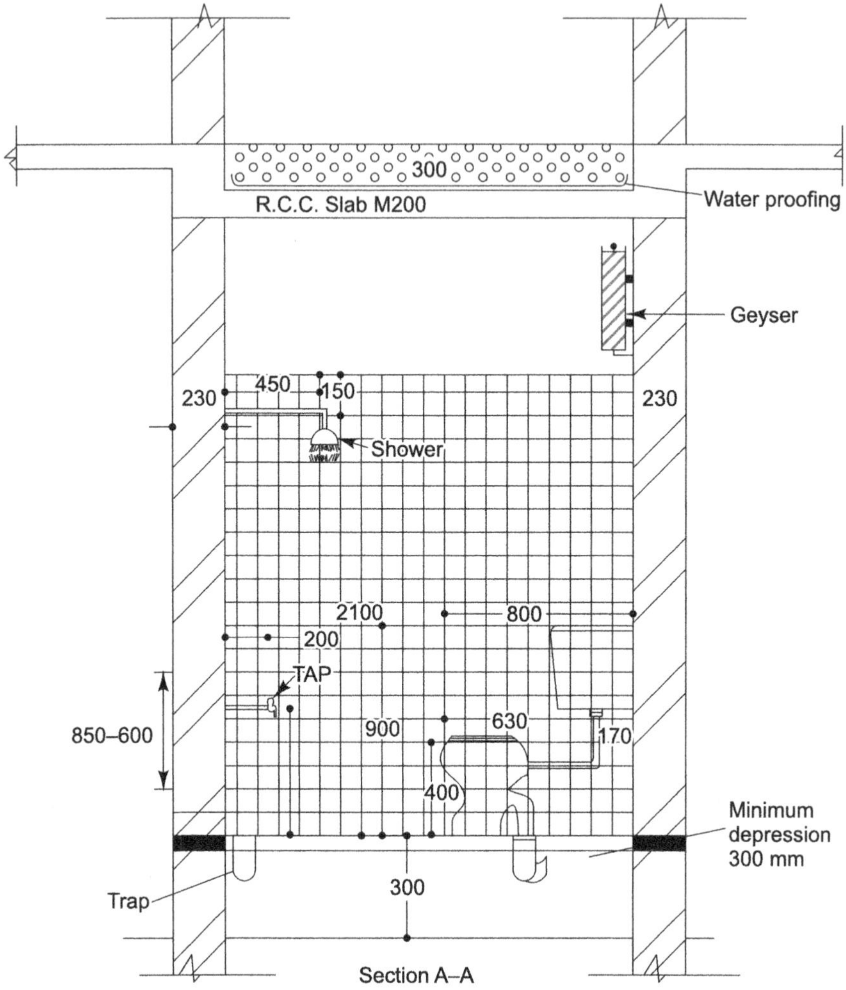

Fig. 3.2 *Toilet details*

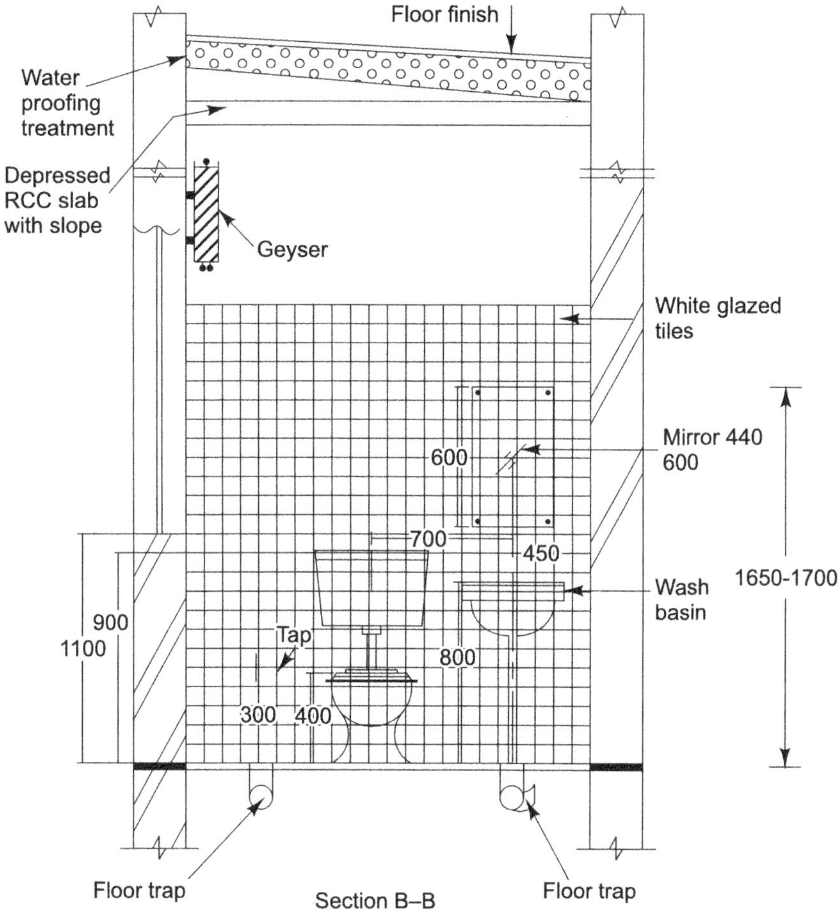

Fig. 3.3 *Toilet details*

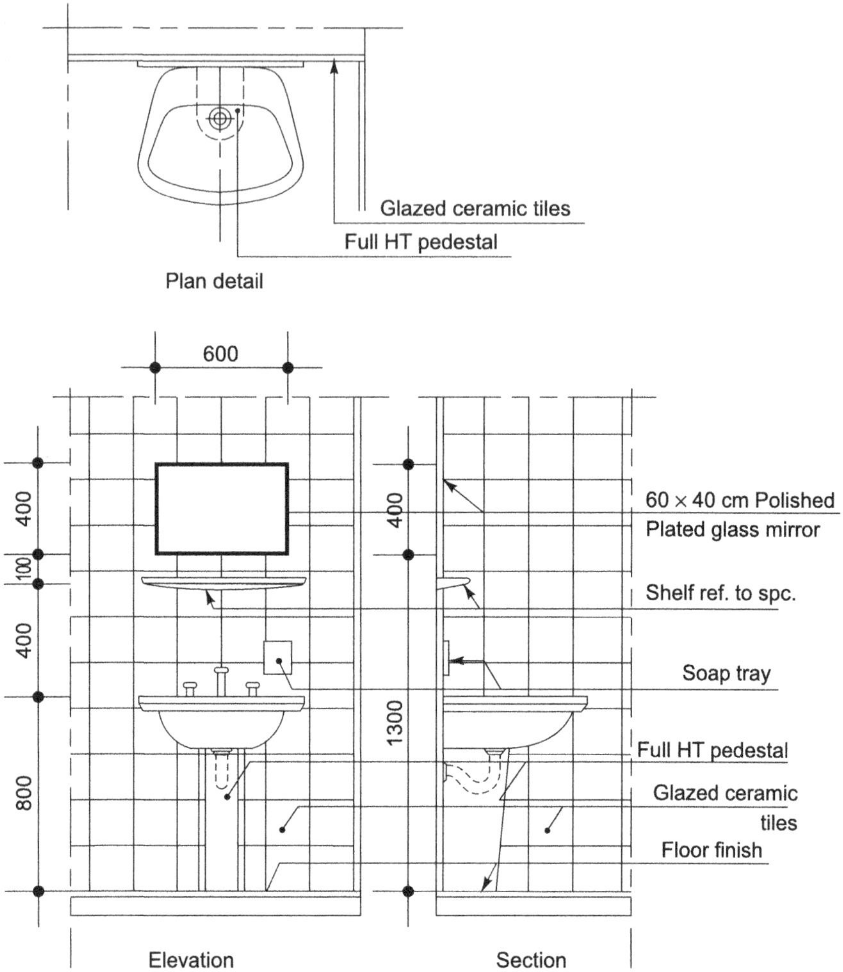

Fig. 3.4 *Lavatory fixing with pedestal*

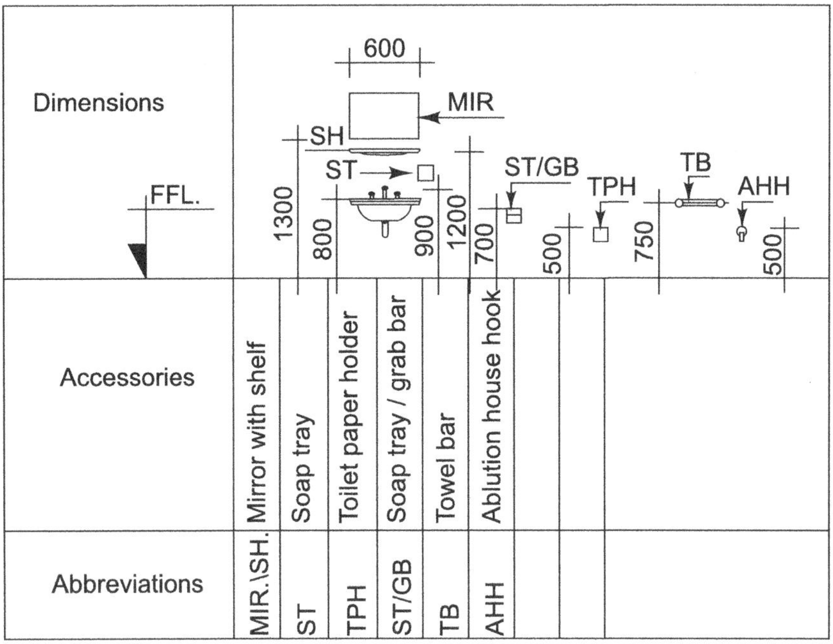

Fig. 3.5 *Toilet accessory schedule*

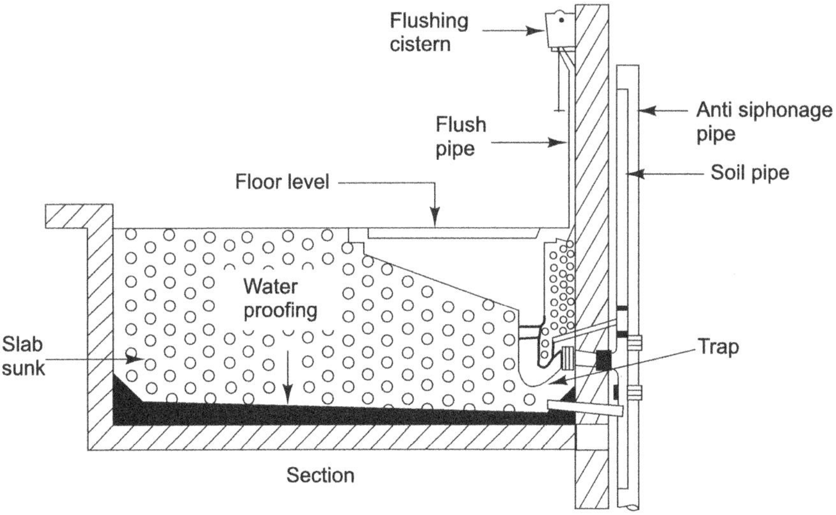

Fig. 3.6 *Toilet details (Contd.)*

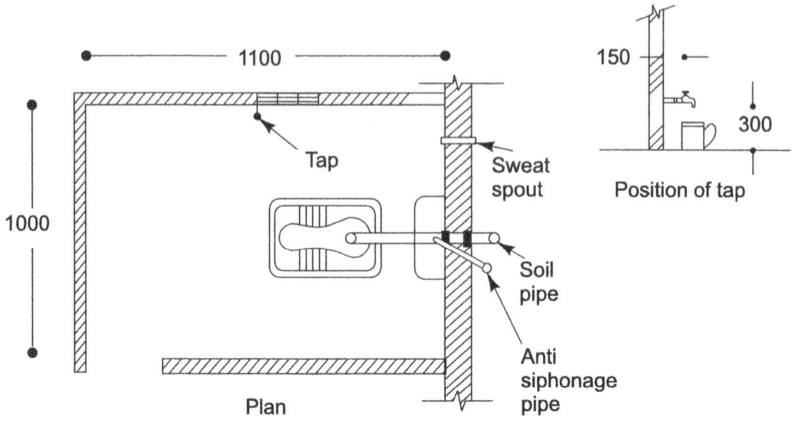

Provision of spout

Fig. 3.6 *Toilet details*

3.2 STRUCTURAL DESIGN

The structural slabs of wet areas should be constructed with a mix of concrete not leaner than M25. Efforts should be to provide slope towards shaft in the top of the surface of the slab. Where services are being accommodated in depressions, minimum area required for accommodating traps/services only should be depressed. It is essential that slab of wet area is made impervious. Structural members required to withstand water pressure should be designed as uncracked sections.

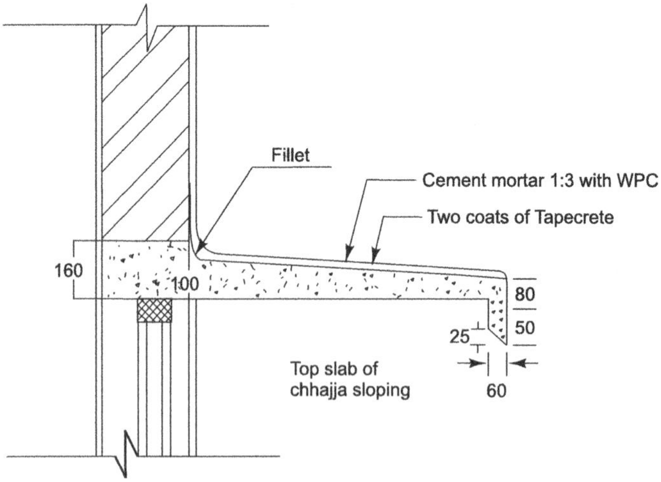

Fig. 3.7 *Water proofing of chhajja*

3.3 GOOD CONCRETE

Concrete is one of the major and most economical building materials known for its long life span and versatility. But concrete, more precisely, cement-concrete is porous in nature. The long life or durability of concrete is greatly dependent on water permeability and ingress of chemicals. In most of the cases, water acts as carrier for harmful chemicals like salts of chlorides, sulphates, alkalis, acids, etc. and these may corrode concrete and/or reinforced steel. Water may also come out from the concrete/ plaster and evaporates from the surface leaving salts and alkalis, which react with paint, and/or making patches. Theoretically, 22–25% (by weight of cement) water is needed to complete the hydration reaction of cement. But in normal conditions, concrete is prepared with extra amount (as much as 45–60% by weight of cement) of water to get desired workability. Any extra mixing water over the minimum requirement of hydration reaction is evaporated from the concrete mass leading to an increase of voids or creating more capillary pores. This is a common factor in all concretes allowing the passage of water and/or water vapour is the presence of inter-connected capillary pores without such voids and their inter connections water or vapour transfer cannot take place.

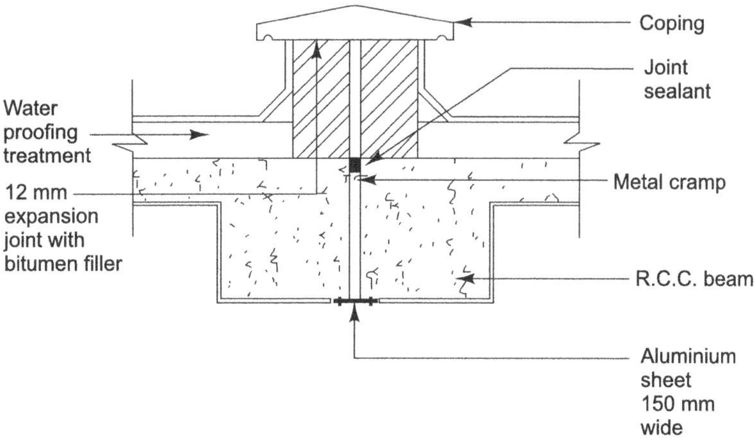

Fig. 3.8 *Raised type expansion joint on terrace*

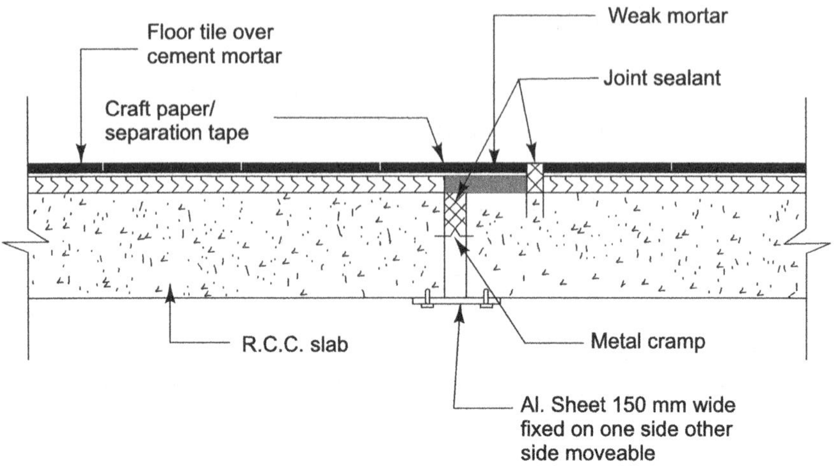

Fig. 3.9 *Expansion joint in floor slab*

Controlling the w/c ratio, using water-reducing admixture can reduce capillary pores. Curing of concrete is also an important factor to continue cement hydration reaction resulting in reduction of capillary pores.

Entrapped voids are created in concrete due to following reasons:

1. Poor mix design.
2. Faulty aggregates gradation.
3. Inadequate compaction.
4. Excessive bleeding and segregation.
5. Very high or very low workability than the requirement.
6. Poor workmanship.
7. IS 456: 2000 PARA 8.2 - Requirements for Durability emphasises that Concrete is more vulnerable to deterioration due to chemical or climatic attack when it is in thin sections, in sections under hydrostatic pressure from one side only, in partially immersed sections and life of the structure can be lengthened by providing extra cover to steel, by chamfering the corners or by using circular cross sections or by using surface coatings which prevent or reduce the ingress of water, carbon dioxide or aggressive chemicals.

As brought out above concrete is a porous material and needs to be treated to improve its water imperviousness. Proper waterproofing of slabs should be done before pipe laying is taken up. In all cases it must be ensured that the waterproofing does not damage while fittings and fixtures are laid. Service designs and their ease of going to shafts should be checked before finalising structural designs so that no structural member is damaged for laying service pipes later. The structural slab/ beam arrangements in a toilet shall be proposed in such a way that there are no inverted beams provided through which the pipe needs to be taken necessitating cutting/puncturing the web of a beam which has to be avoided. In depressed portions a 40 mm diameter pipe should be provided on water proofed slab, projecting into shaft to take away any water seeping down. This pipe also brings to immediate notice the problem of seepage.

3.4 SLOPES

A minimum floor slope of 1:60 should be maintained in floors of wet areas. As far as service pipes are concerned the slopes indicated in Table 3.1 should be maintained.

Table 3.1 Recommended Gradients by "The Institute of Plumbing"

Diameter (mm)	*Surface water drains*		*Foul drains*	
	Main drain	*Branch drain*	*Main drain*	*Branch drain*
	100% Gradient	50% Gradient	80% Gradient	40% Gradient
75	1:75	1:35	1:60	1:30
100	1:100	1:50	1:80	1:40
150	1:150	1:75	1:120	1:60
225	1:225	1:110	1:180	1:90
300	1:300	1:150	1:240	1:120

Source: Plumbing Engineering Services Design Guide.

3.4.1 Slopes in Structural Members

(i) To reduce load on structural slab, provide slope (1:100) towards shaft in the top of the surface of the structural slab itself.

3.5 DESIGN OF SERVICES

(a) All services should be designed well in advance in accordance with guide lines issued by The Institute of plumbing or codes of practice as applicable and coordinated drawings prepared.

(b) Underground piping should be laid at a depth where it is not affected by frost or traffic movements.

(c) In flats and apartments where water supply is from a common service pipe, a stop valve must be fixed on branch pipes to control groups of ball valves and draw off taps so as to minimise interruption of water supply and facilitate maintenance works.

(d) Water for general use, drinking and cooking purposes should be, to the extent possible, on branch pipes directly connected to service pipes.

(e) When a pipe is required to pass through a wall or floor, a sleeve shall be fixed for insertion of the pipe and allowing expansion/contraction.

(f) Services drawings should be issued to site staff well before execution of the work.

3.6 DRAINAGE SCHEME

(a) The site drainage is to be planned first. It must be ensured that the whole area is properly drained with network of storm water drains.

(b) Terrace and court yard drainage must be planned at the planning stage. One should avoid long run of water on terrace and ensure adequate slopes.

3.7 RECORDS OF DESIGN AND SERVICES DRAWINGS

All service lines and fittings/fixtures and their disposal system should be marked on drawings and these drawings should invariably be kept safely with completion plans as part of maintenance manual. One copy should be available with the maintenance engineer/agency for ease of reference.

3.8 ACCESSIBILITY FOR MAINTENANCE

While planning services it is essential to keep the maintenance of these services in mind. There should be space around the pipe for fixing wrench and working on it. Vent doors and pipes should be available for inspection and easily approachable. Adequate clean outs should be provided.

For requirements of platforms required to facilitate maintenance refer section 9.6

CHAPTER 4

MATERIALS

4.1 CHOICE OF MATERIAL

Materials to be used in the wet areas need special attention. Materials used for floors and walls have to be pleasant looking, non-absorbent, durable and should be smooth but not slippery, not having fouling surfaces. Needless to say that fittings and fixture should be of durable material, free from defects, pin holes, etc., and be compatible with piping materials used. These should be preferably tested to be sure of quality. Materials of reputed firms/manufacturers should only be used. Careful selection of materials for external and internal works is suggested. Materials chosen have to have:

1. Smooth bore
2. Durability and having long life
3. Easy to maintain and future repairs.
4. Ease of availability
5. Non-corrosive and not susceptible to Bacterial/fungus growth
6. UV resistant
7. Smooth and easy to work with
8. Capable of taking pressure and temperature of water/fluids

4.1 PIPES

So far Cast Iron pipes have been used for water supply for all external works and drainage. SW and RCC hume pipes were used for sewerage and storm water drainage in external areas. For domestic drainage works CI/GI pipes were in use till a decade back which is now getting replaced by PVC pipes of different categories. Now-a-days for house drainage choice following materials in pipes are available:

- Cast Iron
- Copper
- Galvanised Steel
- HDPE (High Density polyethylene)
- MDPE (Medium Density polyethylene)
- PVC (Polyvinyl chloride)
- UPVC (Unplasticised Polyvinyl chloride)
- PPR (Polypropylene Random Copolymer)
- CPVC (chlorinated polyvinyl chloride)

For internal plumbing work, materials have been discussed in greater detail in subsequent paragraphs, and relative merits and demerits have been brought out to facilitate user in making choice based on available facts.

4.2 PIPE MATERIALS FOR PLUMBING

In old construction, lead plumbing was common. It got generally replaced by galvanised iron water pipes which were attached with threaded pipe fittings. Higher durability and costly systems were made with brass pipe and fittings. Copper with soldered fittings, though had been used as early as 1900, but became popular around 1950, In Copper pipe plumbing bacteria can't grow and is bacteriostatic.

Plastic supply pipes have become increasingly common since about 1970, with a variety of materials and fittings employed, however, plastic water pipes do not keep water as clean as copper and brass piping does.

4.3 GALVANISED STEEL PIPES

Galvanised steel potable water supply and distribution pipes are commonly found with nominal diameters from 3/8" to 2". Galvanised steel is relatively expensive, difficult to work with due to weight and requirement of a pipe threader. It remains in common use for repair of existing "galv" systems and to satisfy building code non-combustibility requirements typically found in hotels, apartment buildings and other commercial applications. It is also extremely durable. Black lacquered steel pipe is the most widely used pipe material for fire sprinklers and natural gas.

In addition to being expensive, GI pipes have a tendency to obstruct due to internal rusting and mineral deposits forming on the inside of the pipe over time after the internal galvanizing zinc coating gets degraded. In potable water distribution service, galvanised steel pipe has a service life of about 30–50 years, although it is not uncommon for it to be less in geographic areas with corrosive water contaminants.

Copper tubing is most often used for supply of hot and cold water, and as refrigerant line in HVAC systems. There are two basic types of copper tubing, soft copper and rigid copper. Copper offers a high level of resistance to corrosion, but is becoming very costly.

4.4 PLASTIC PIPES

Plastic pipe is in wide use for domestic water supply and drainage, waste and vent pipe.

4.4.1 PVC—Polyvinyl Chloride

- strong and rigid
- resistant to a variety of acids and bases
- may be damaged by some solvents and chlorinated hydrocarbons
- maximum usable temperature 140°F (60°C)
- usable for water, gas and drainage systems
- not useable in hot water systems

PVC pipes can be used for cold water only, or venting. CPVC can be used for hot and cold potable water supply.

4.4.2 UPVC Pipes

The density of UPVC is almost one sixth the weight of cast iron and steel, thus making it much cheaper to transport and easier to handle during installation. It has good mechanical strength, and hence is suitable even in varying conditions.

It is highly capable of fighting attacks by fungi and is not subject to contamination. The inside surface, which is extremely smooth, does not support any growth, encrustation, or fuming, and no odour or taste is transmitted to the fluid being transported.

Being a thermoplastic material, UPVC is more capable to withstand deformation.

UPVC pipes do not support combustion and are self-extinguishing. Therefore, they are ideally suited for use in buildings and other constructions.

PVC pipes require closer supports, and such supports should allow repeated temperature movement.

4.4.3 CPVC—Chlorinated Polyvinyl Chloride

Similar to PVC but designed for water up to approx. 180°F (82°C), chlorinated polyvinyl chloride (CPVC) is a thermoplastic pipe material. CPVC pipe has a number of features that make it an improvement over standard PVC piping. It offers greater heat resistance withstanding corrosive water temperatures. CPVC is non-toxic, while PVC may leech toxins into water at increased temperatures. CPVC also offers greater strength and flexibility, while PVC is far less ductile.

CPVC pipe has become a modern standard for water supply and liquid lines because of its cost saving benefits over metal piping.

CPVC pipe is immune to galvanic corrosion and resists scale build up. It is also resistant to chemicals, and durable against its residues. Installation is safer and faster, requiring only special solvent cement. CPVC piping generally lasts longer, as it is less susceptible to failures.

Some people complain of a plastic taste in the water. The pipe and fittings are subject to cracking if dropped or during earthquakes, and can expand with temperature change. Bacteria can grow inside the pipe as well.

CPVC piping is recognised in all model plumbing codes. CPVC is fire resistant and will not burn without a flame source, making it suitable for fire suppression systems in light hazard and residential settings.

4.4.4 PPR Pipes

PPR pipes are manufactured from Polypropylene Random Copolymer raw material. Polypropylene-random (PP-R) water supplying pipes and fittings are a kind of green building materials, which can be used in hot water systems besides being used in sanitary networks. Socket welding is used as the connection method. The production is between Ø 20 and Ø 315 mm.

It has a long service life; the piping system can be used for more than 50 years. It is non-toxic, heat-resistant, durable, light weight, and has smooth inner walls that reduce pressure loss and increase flow speed. It offers no calcification, heat preservation and energy saving, low laying time and cost, natural taste and odour and good resistance to chemicals.

4.5 PPR PIPE FITTING INSTALLATION

Installation of PPR pipes is done by Fusion method. Following steps are involved:

1. Mark the pipe

   Mark the pipe for depth of penetration into the heater bush and fitting. The mark must remain visible under heating and joining.

2. Heat the pipe and the fitting

   Push the pipe and the fitting into the heating tools. Once pipe and fitting are hot (after the correct time), pull out pipe and fitting very slowly.

3. Joint the pipe and the fitting

   Joint the pipe and the fitting, and push the pipe until it reaches the mark (that has to stay outside). During jointing, the welded part of pipe and fitting must remain fix, without any rotation. During the cooling time, the welded part of pipe and fitting can be adjusted until cold.

4. Fusion inspection

   The outer fusion seam must be inspected. The seam must be present all around the pipe.

4.6 S.W.R. PIPES

SWR pipes conform to IS-13592 specifications, and fittings conform to IS-14735 specifications, SWR pipes and fittings are best preferred for drainage system for homes, offices, hotels, commercial complexes, etc. These pipes are available in 75–160 mm sizes. Salient features are:

- Strong and durable
- Easy to handle and join
- Better flow characteristic
- Leak proof
- Energy saving
- Safe
- Resistant to rust, UV radiation and most of the chemical actions
- Economical
- Maintenance free
- Self coloured
- Better strength to weight ratio

4.7 COMPARISION OF VARIOUS PIPING MATERIALS

Weight and thickness of PPR or CPVC Pipes to be used should be checked and category of pipe be selected depending upon the use. Also skill of personnel available to carry out the work should be gone into before installation work is entrusted. Proper trained staff is necessary to carry out correct jointing and installations for PPR pipes to a greater extent as controlled fusion work has to be done. Even for CPVC piping trained personnel only must be engaged.

The comparison is given in Table 4.1.

Table 4.1 Comparison of various piping materials

Criteria	*GI Pipe*	*Copper Pipe*	*HDPE Pipe*	*PVC Pipe*	*PP-R Pipe*
Effect of hard water	High scale formation	Scale formation is prohibited due to smooth bore	Scale formation is prohibited due to smooth bore	Scale formation is prohibited due to smooth bore	Scale formation is prohibited due to smooth bore
Effect of soft water	Gets corroded	Gets corroded due to acidic nature of water	No effect	No effect	No effect
Health criterion	Low due to lead content and corrosion	Good with ferrule but lead content in solder is bad for health	Very good	Very good	Very good
Jointing techniques	Threaded	Soldered/ Ferrule	Fusion weld	Solvent cement	Fusion weld
Corrosion resistance	Very low	Low	No effect	No effect	No effect

Criteria	*GI Pipe*	*Copper Pipe*	*HDPE Pipe*	*PVC Pipe*	*PP-R Pipe*
Thermal strength property at 60°C temp.	Very good	Very good	Limited	Not recommended	Very good
Availability of fittings	Very good	Average	Low	Good	Very good
Thermal expansion	Low, good for concealed piping	Low, good for concealed piping	Very high, not to be used for concealed piping	Very special care is required for concealed piping	Low, good for concealed piping
Effect of sub-zero temp.	Up to 0°C	Up to 0°C	Up to −40°C	Up to 0°C	Up to −40°C
UV resistance	Very good	Very good	Very good	Low	Very good
Ease in installation	Low	Average	Low	Good	Very good
Flow properites for friction	Low	Very good	Very good	Very good	Very good

CHAPTER 5

FITTINGS AND FIXTURES

5.1 FITTING MATERIALS

The bodies of fittings for pipe and tubing are most often of the same base material as the pipe or tubing being connected. However, any material that is allowed by code may be used, but must be compatible with the other materials in the system, the fluids being transported, and the temperatures and pressures inside and outside of the system.

Potable water supply systems require not only pipe, but also many fittings and valves which add considerably to their functionality as well as cost.

5.2 FITTINGS AND VALVES

Fittings and valves are used in pipe and plumbing systems to connect straight pipe or tubing sections, to adapt to different sizes or shapes, and to regulate fluid flow. Fittings, especially uncommon types, can be expensive, and require time, materials and tools to install.

So this aspect should be kept in mind while deciding on piping and plumbing systems.

Some of the items need special care to ensure efficient running of the system. A few of these are described below:

5.2.1 Floor Trap

To prevent ingress of foul air from sanitary network, a water seal normally of 50 mm, is required in floor traps. Trap allows flow of liquid with suspended soil but prevents flow of air on account of its water seal. It must be self cleansing type and not allow loss of its water seal.

Defective Installation is the main cause of leakage. The defective floor installations can be summarised as:

- Installation where trap is seated too deep from the floor level;
- Installation where one or more pipe enters into floor trap;
- Installation where trap is over flooded.

The level of trap is determined by the level of connection of the horizontal piping. The present practice is to locate the trap below the floor and seat it into a mass of concrete. Seating of trap on concrete alone may not stop leakage. Therefore, a watertight lead joint should be provided up to the floor level in CI trap.

Wrong fixing of floor traps normally results in leakage/seepages. Following precautions would help in eliminating this problem.

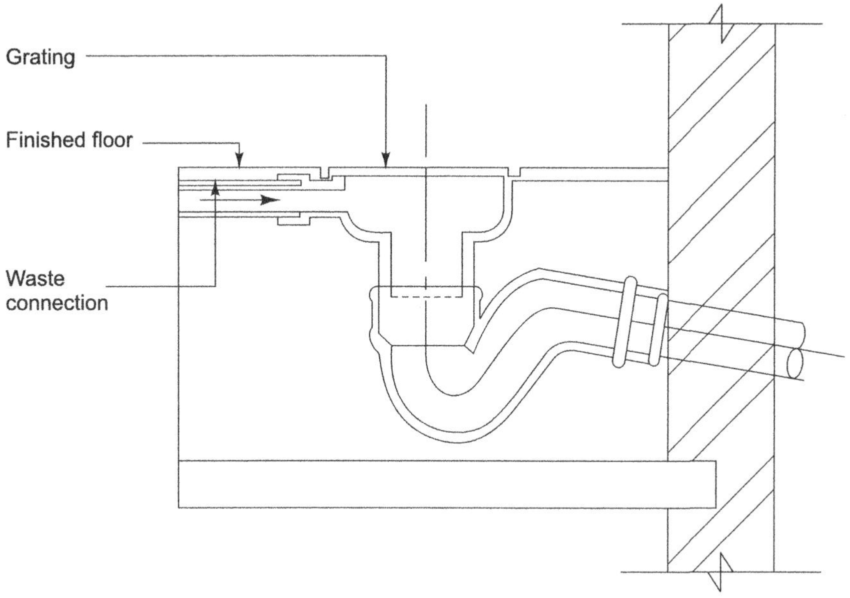

Fig. 5.1 *Trap with inlet fitting*

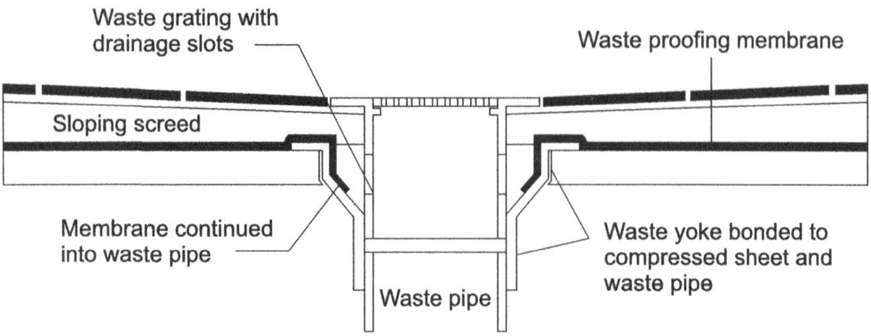

Fig. 5.2 *Waste detail*

(i) Design services so that only one pipe discharges into trap. Details to be used for disposal of pipe are shown in Fig. 5.1. The gap between floor surface and top of traps many times results in water not discharging directly in the trap. If the detail at Fig. 5.2 is followed, this problem would be avoided. Sides of trap should be done with rich concrete with no gap.

(ii) Pipe(s) discharging on to trap should project in to the trap but not cause obstruction and should not be short of length. (Fig. 5.3)

(iii) Cover of trap should be designed such that it cannot be removed from the trap by careless cleaners.

(iv) Trap should be as per IS: 3989.

(v) A well defined waterproof depression should be provided around the trap with proper waterproofing to ease out excess water.

(a)

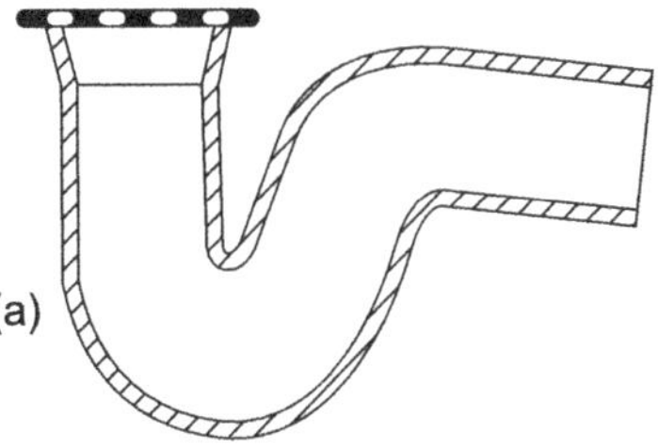

(a) Correct type of floor trap.

(b)

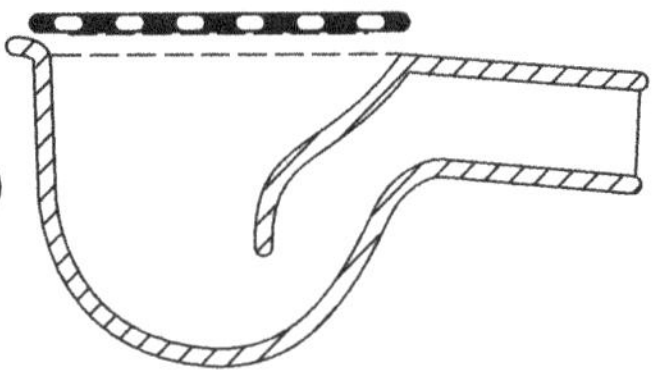

(b) In-correct type of floor trap-protrusion type. This is not self-cleaning and has projection where it can get clogged.

(c)

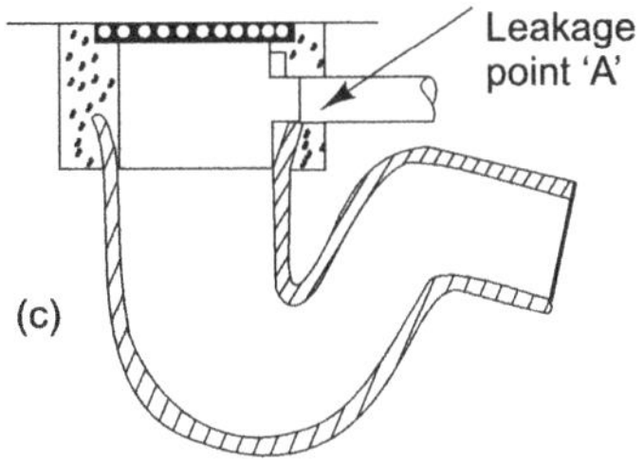

(c) When a waste pipe discharges into the floor trap below floor level the junction is made of cement, concrete. Point 'A' below the pipe is vulnerable as a leakage point due to improper compaction of concrete below pipe.

(d)

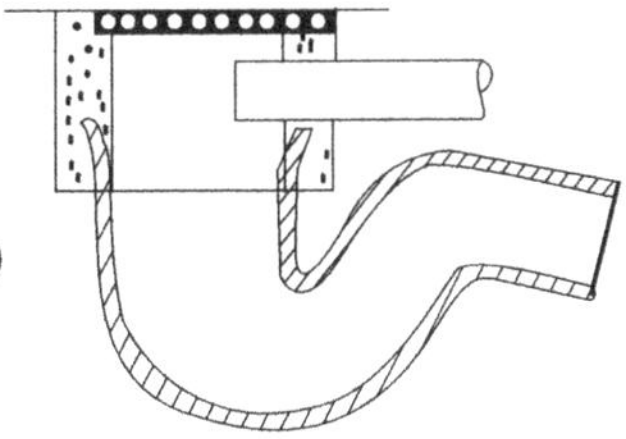

(d) More often the waste pipe is not cut to the exact size, it may be short as in Fig. 5.3(c) or over shoot the edge as in Fig. 5.3(d) which again causes flooding of water and water may find its way into the floor filling below.

(e)

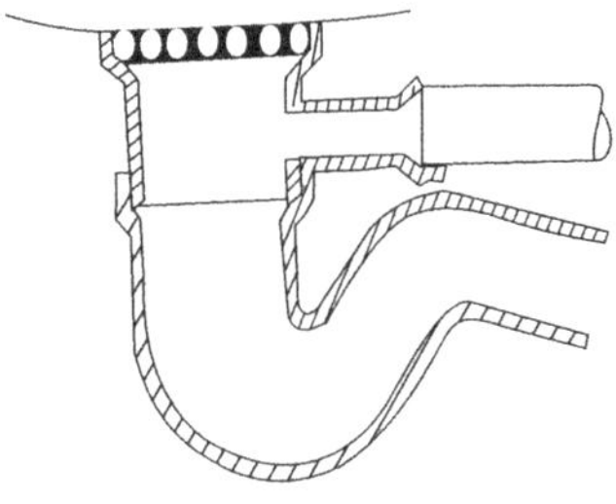

(e) The correct method of making these joints is to have a G.I. or M.S. custom built 'T' junction with lead caulked joints.

Fig. 5.3 *Floor traps and jointing*

5.2.2 Self Sealing Waste Valves

Waste valves replace the water seal trap and are a recent introduction. It reduces pipe work whilst meeting the mandatory codal requirements. These valves open to allow the flow of water from the appliance, or to allow air to enter the pipe network system in case of negative pressure, and then close automatically when the flow stops or the pipe work network reaches equilibrium with atmosphere. Thus the system is self ventilating. These valves are particularly useful for situations where seals could be lost by evaporation.

5.2.3 Sanitary Pipes

As already mentioned, pipes of only very good quality should be used, duly tested for no defects. Joints should be minimum. Adequate slope is must to ensure proper flow. Proper clean outs should be provided. When bends are provided, they may be of larger radius. Pipes should not be concealed in masonry/concrete/walls, etc., to the extent possible.

5.2.4 Water Closets and Flushing Cisterns

High level flushing cisterns have given way to low level cisterns. Cisterns which have water saving facilities by allowing controlled flushings, should be preferably used to save the now depleting resources of water. Special attention need be paid to the flushing pipe and WC connection. The pipe must go into the connecting hole of WC. When Indian WC is used, we may preferably use G.I. pipe duly bent hydraulically to shape for flushing. Pipe may be given protective coat of bitumastic painting. Joints should be waterproofed. All working parts of cistern should be designed to operate smoothly and efficiently. Cistern should be mosquito proof.

Location of WC should be as per dimensions shown in standard details. Many a times, it is seen that the lid of seat cover keeps falling as the distance between the wall and WC has not been kept properly, resulting in the cistern not allowing vertical stay of cover. Breadth of the low level cistern should be taken into account to ensure that cover or seat of water closet come to rest in a stable position when raised (Fig. 3.1).

Joint of WC and trap (P or S type) gives its own problems due to the dissimilarity of materials. This joint should be made using nylon or rubber "O" ring. Use of cement mortar 1:1 with non-shrink waterproofing compound should be made in the joint.

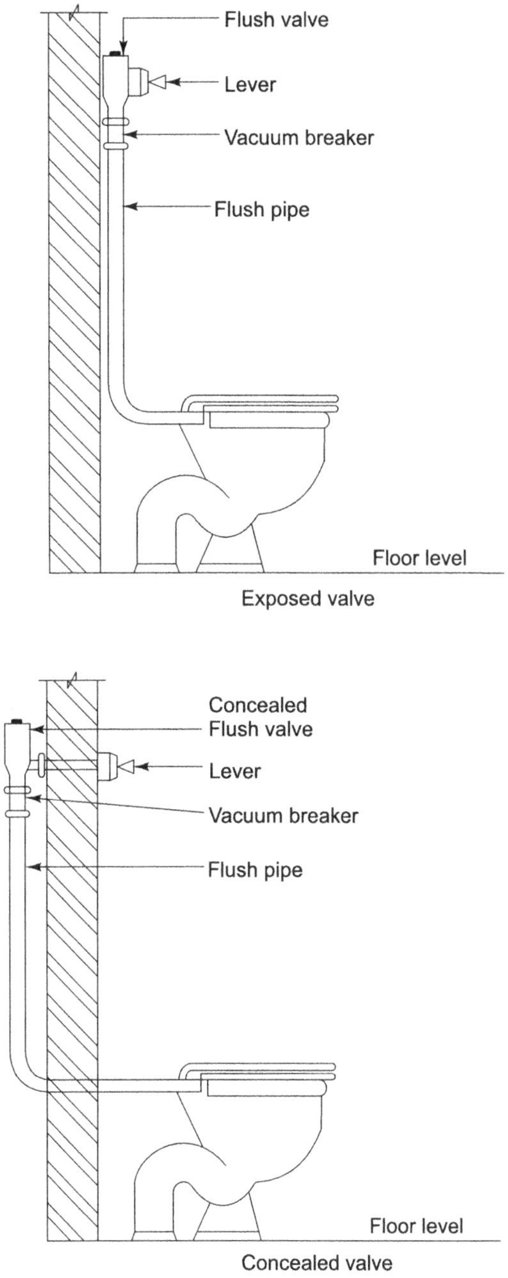

Fig. 5.4 *Flush valve*

Instead of using WC with separate foot rests it is better to use "Orissa pattern" integrated WC. First, temporary laying of pipe line and WC pan should be done by giving temporary packing. The level should be checked and corrected taking into account the final finish level of flooring. The WC pan may be taken out and pipe joints filled up. After WC pan is fitted and jointed, cement concrete block should be put around this joint and then depressed portion filled up.

Overflow pipes from cisterns should be brought down to floor levels.

5.2.5 Fixing of European Type WC

It is important to ensure that WC is fixed to floor with the help of strong screws which are properly tightened. European WC should be fixed to floor with screws using rubber gasket to serve as a cushion and water barrier. Junction between pan pedestal with flooring should be filled with cement mortar mixed with non-shrink waterproofing compound (Fig. 5.5).

WC's should not be provided at a level that they become susceptible to back flow.

Fixation of urinals and the dimensions to be specifically taken care of are drawn in Figs. 5.6–5.8. If these installation are fixed properly these would ensure ease of use and privacy.

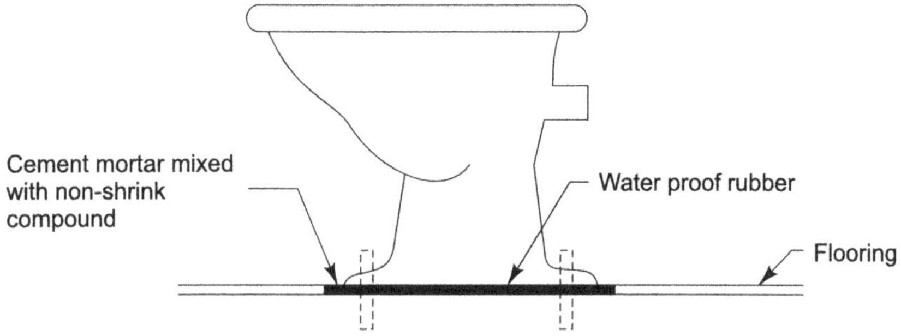

Fig. 5.5 *Fixing of European type water closets*

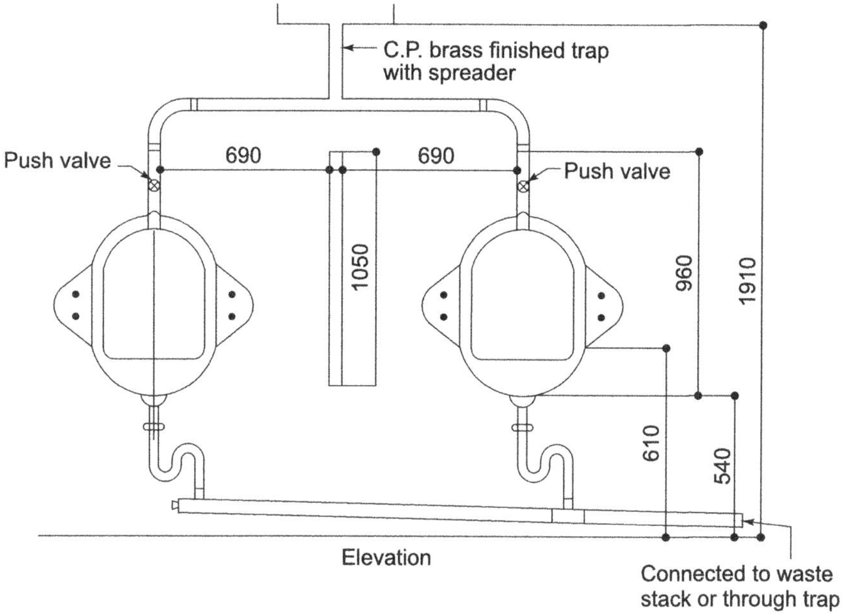

Fig. 5.6 *Urinal*

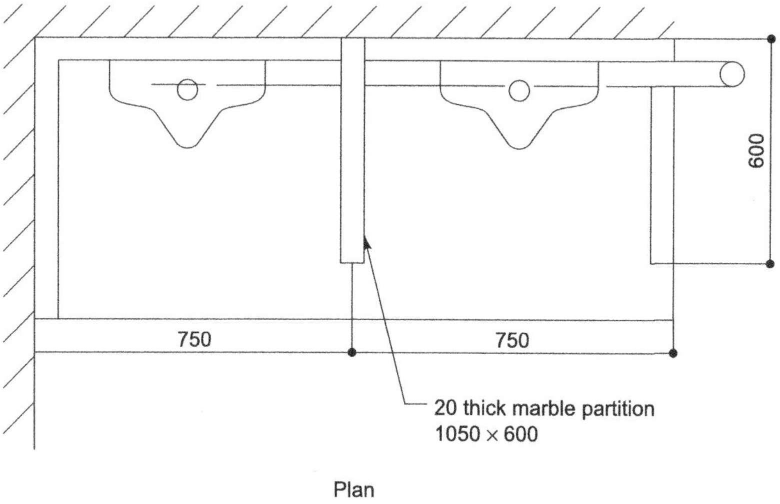

Fig. 5.7 *Urinal*

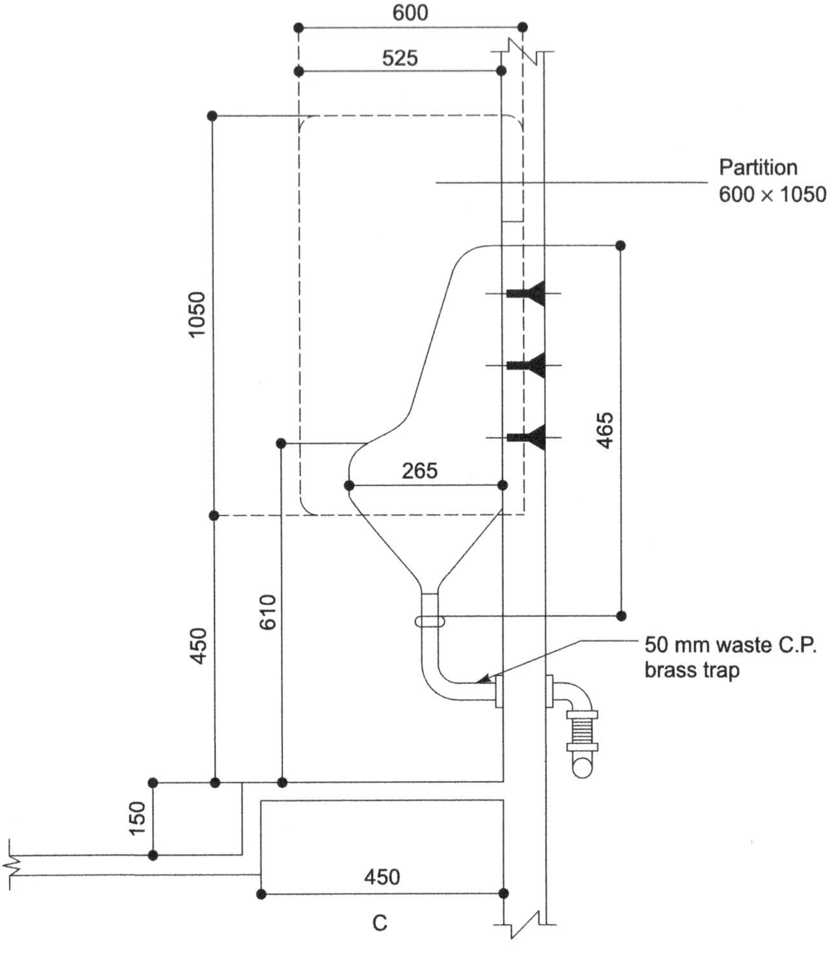

Fig. 5.8 *Toilet details*

CHAPTER 6

EXECUTION STAGE

Having given due attention to planning and design details with prevention of leakages and seepages in mind, next obvious important part is its implementation. During execution stage, we have to ensure adoption of correct methodology, sequence of operation and attention to minor details to achieve our objectives. Well-planned scheme will go haywire unless it is implemented with precision and care it requires.

6.1 GETTING STARTED

(a) While works of wet areas are taken up, shop drawings for each area based on the architectural planning should be prepared by executing agency and approved by the project in charge.

(b) The sizes of tiles or stones are to be decided keeping the size of the area to be done up so that joints are minimum and follow an aesthetic pattern.

(c) Cutting of tiles should be minimised by pre-arranging on paper the location of WC's wash-basins and other appliances and fixtures.

(d) While fixing G.I. or UPVC pipe in chases these should be securely clamped specially at bends and joints of fixtures. Maximum vibrations take places at bends and joints and hence proper clamping at these locations. Minimise vibrations caused due to repeated use and reduce chances of leakage/seepage.

6.2 BASIC STEPS

a) Plans and elevations of the floors and walls should be prepared with fixtures clearly marked and can be harmonised with the pattern of proposed tiles. Location of each plumbing and sanitary fixture, light fixtures, geyser, sockets and switches, etc., shall also be marked on drawings before execution is taken up. For ensuring proper locations guidance can be taken from the standard dimensions given in Figs. 3.4 and 3.5.

(b) Special care needs to be taken to see that the electric outlets are not located in wet areas.

(c) Appliances and pipes are to be so arranged so as to allow close grouping of connections near the main soil pipe(s). No jointing material should project inside the pipe.

(d) Clear gap between wall and soil/drainage pipes helps in avoiding dampness.

(e) To prevent dampness travelling through capillary action, DPC should invariably be provided. DPC should be provided continuously below doors and extended below floors as shown in Fig. 6.1. Where DPC has not been provided or has failed, this can now be introduced using techniques discussed in waterproofing section.

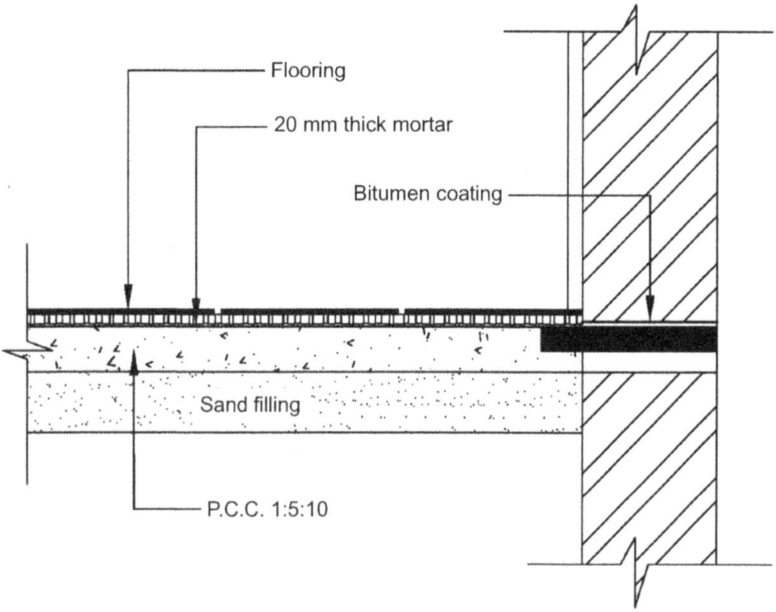

Fig. 6.1 *DPC laying*

6.3 PRECAUTIONS

Attention to some minor details during execution goes a long way in ensuring proper trouble-free installations. Some such points, which one needs to keep in mind, are:

(a) Provision of unions in pipes near stop cocks.

(b) Making flush pipes secure to cistern outlets using coupling net.

(c) Protecting GI pipes with bitumen painting and covering where embedded in concrete or walls.

(d) Floors are made joint less or joints are effectively sealed. Floors are provided with adequate slopes towards traps.

(e) Wet area walls are made of impervious materials to a height where water splashes are likely. In shower areas it could go up to 1.5 – 2 m height.

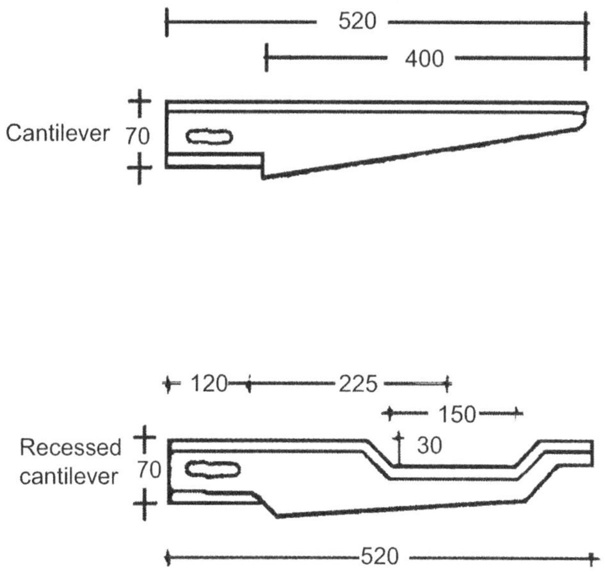

Fig. 6.2 *Bracket for sink*

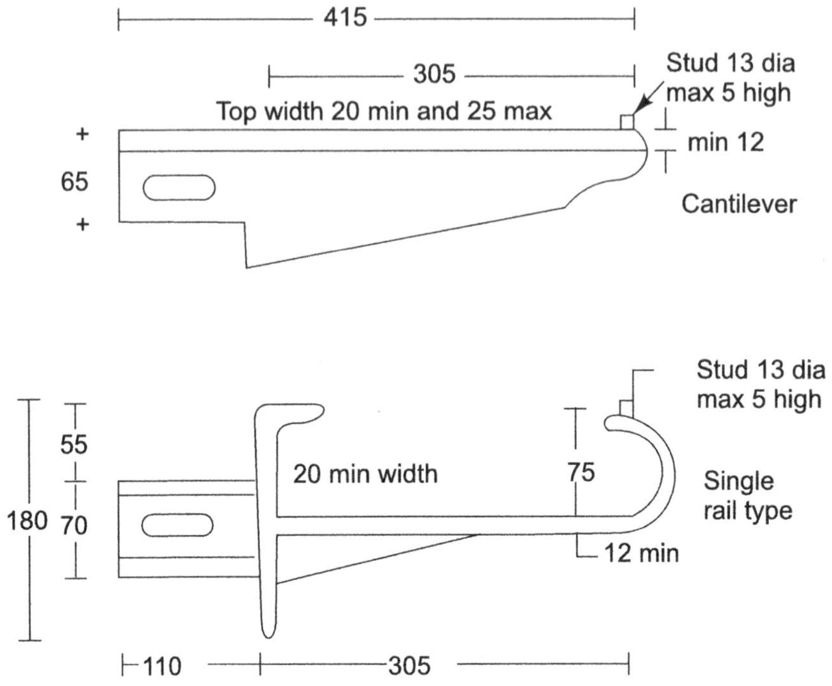

Fig. 6.3 *Bracket for wash basin*

(f) Proper quantity of lead should be used in lead caulked joints. Yarn packing should be limited to two rounds only and it should be properly packed.

(g) It must be ensured that the taps are fitted by protruding the pipeline at least by 20 mm beyond the finished surface of dado to avoid leakage/dampness on the wall besides easy operation of taps. Gaps between wall and pipes must be sealed.

(h) In case of cement mortar joints between WC pan and trap, proper curing is done before all round filling is carried out.

(i) PPR pipes when used, should be fusion welded by skilled persons to avoid weak/improper joints.

(j) Provide grab rails alongside WC, Bath, shower and basin for use of elderly and should be positioned so that accidental misuse is avoided.

(k) The tap closest to bath should be cold tap for safety of children.

(l) Cast iron brackets when used should confirm to IS:775 and should be properly protected by suitable impervious paints. M.S. angles or Tee sections as brackets should not be allowed. Bracket shall be fixed in position first and then only dado work should be taken up. (See Fig. 6.2 and 6.3).

6.4 JOINTING REQUIREMENTS

6.4.1 Weight of Pipes

Sand Cast Iron (S.C.I.) pipes confirming to IS:1729 and spun cast iron pipes confirming to IS:3989 shall be used for sanitary installations. These should be purchased from reputed dealers to avoid any sub-standard materials. Weight of pipes can be checked. For S.C. I. pipes for a length of 1.8 m following weight is acceptable.

Sl. No.	Diameter	Weight
1.	50 mm	11.41 kg excluding ears 16.52 kg
2.	75 mm	16.52 kg excluding ears
3.	100 mm	21.67 kg excluding ears

For spun pipes the weights for 1.8 m pipe shall be:

Sl. No.	Diameter	Weight
1.	50 mm	8.40 kg
2.	75 mm	12.50 kg
3.	100 mm	18.80 kg

6.4.2 Lead Caulking

The lead used for caulking shall be soft bluish grey pig lead free from admixtures of tin or other impurities. The spun yarn shall be clean hemp and soaked in hot tar or bitumen, cooled and dried before use. While jointing the pipe, the interior of socket and exterior of spigot should be thoroughly cleaned and dried. The spigot end should be inserted into the socket right up to the back of the socket and carefully centred by 2–3 laps of treated spun yarn twisted into coils of uniform thickness, well caulked into the back of the socket. The individual piece of yarn should not be less than the circumference of the pipe. Lead is to be poured later and well caulked for a tight fit. Workmen have a tendency to utilise more of yarn than necessary and less of lead. Caulking of joints is also not done. This is required to be checked.

Pipe should give a ringing sound when struck with a light hammer. S.C.I. pipes should be painted internally with Dr. Angus Smith Solution.

Record of the lead used in the joints should be maintained. Following quantities of lead have been found to give sound joints.

(a) For water supply lines:

80 mm diameter – 1.80 kg

100 mm diameter – 2.20 kg

125 mm diameter – 2.60 kg

150 mm dia – 3.40 kg

(b) For Rain water pipes:

50 mm diameter – 0.50 kg

75 mm diameter – 0.66 kg.

100 mm diameter – 1.00 kg.

For sanitary pipes: (Where hidden or under floors)

50 mm diameter	-	0.77 kg
75 mm diameter	-	0.88 kg
100 mm diameter	-	0.99 kg

6.5 GENERAL PRINCIPLES OF PLUMBING/SERVICES LAYING

(i) Stoneware pipes should not be carried through foundation walls as these pipes are liable to be broken by uneven settlement.

(ii) No pipe carrying sewage should be buried under the basement floors. It should be placed in masonry-lined trenches with removable covers.

(iii) Sanitary fixtures should be so fixed as to be completely exposed to view and easy to clean. As far as possible, sinks and water-closets should be installed against outside walls.

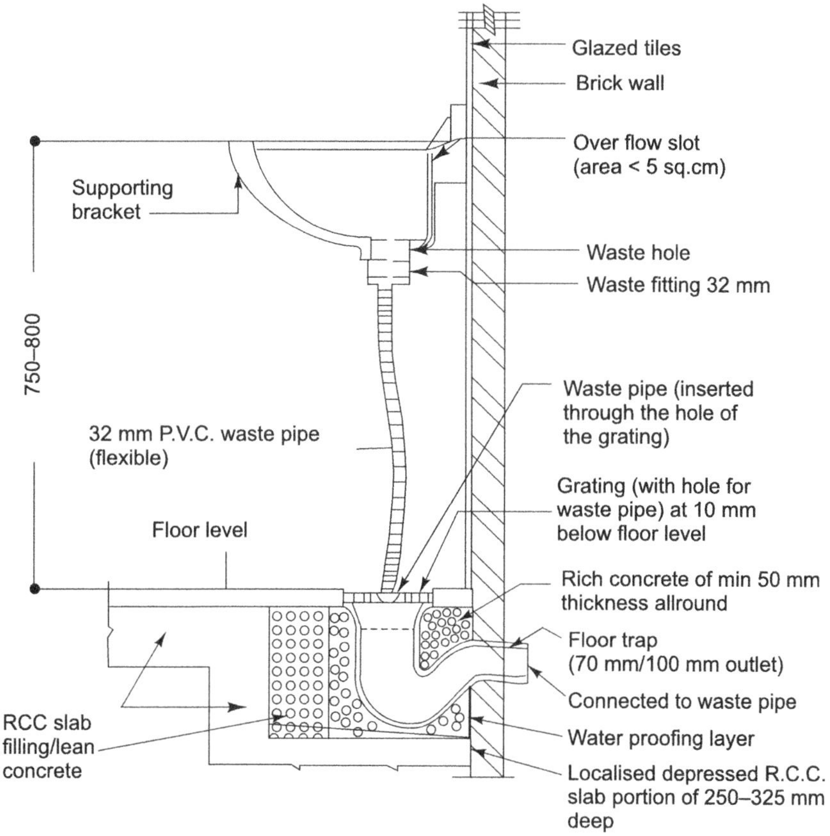

Notes:

Recommended for quarters and in general offices. (Drawings not to scale, All dimensions are in mm)

(Courtesy CPWD Specifications)

Fig. 6.4 *Toilet details wash basin fixing (open waste pipe)*

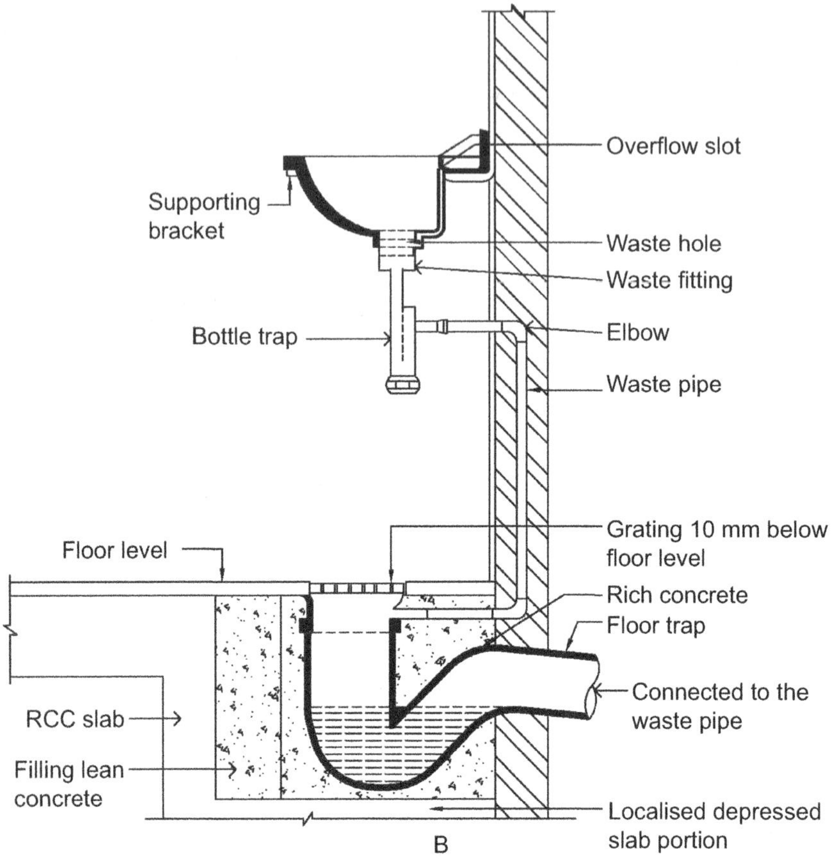

Notes:

Recommended for higher type of quarters, bungalows, five star hotels and for personalised use.

(Drawing not to scale, All Dimensions are in mm)

(*From CPWD Specifications*)

Fig. 6.5 *Toilet details*

(iv) Where wash basins are fixed on platform (e.g., for public use), It is preferable to provide white glazed floor channels to receive the wastes from series of wash basins, so as to avoid the necessity for numerous traps and complicated pipe work. These channels to discharge into the waste pipe through 10 cm × 5 cm bell-mouthed floor traps provided with gratings. (See Figs 6.4 and 6.5)

(v) No soil, waste or vent pipes shall be drilled or trapped.

(vi) The drainage and vent pipes shall be kept free of all other services, such as drinking water. No soil, waste or vent pipe shall have any rain water pipe connected to it. (See Fig. 6.6)

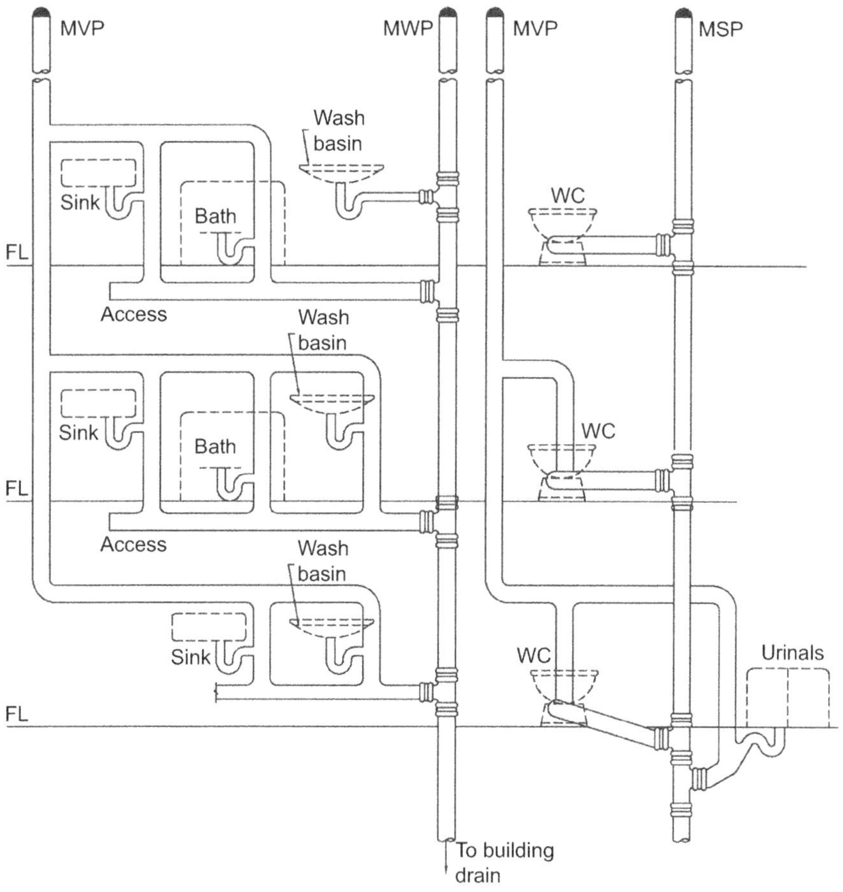

Fig. 6.6 *Diagram of two pipe system*

6.6 FLOORING

Flooring of the wet areas should consist of hard materials that are non-absorbent. Joints in floors should be kept to absolute minimum. With this in view it is better to provide cast *in situ* Terrazzo flooring instead of Terrazzo tiles in the wet areas. Use of Kota or marble stones is desirable. When stone flooring is provided they may be laid in predetermined shape and sizes especially in corners of the rooms so as to facilitate the two directional slopes in flooring.

Now-a-days there is a range of vitrified tiles and ceramic available with variety of colour combinations. These should be used with care. Proper bedding should be prepared and joints sealed effectively. To ensure that no water percolates below, bed of waterproofing mortar must be used.

Joint between the floor and wall is always a weak point. Flooring material should invariably go below the skirting or dado finish up to the wall surface so that no water goes below the flooring through joints. Adequate slopes in pre-designed fashion must be ensured. (See Fig. 6.7)

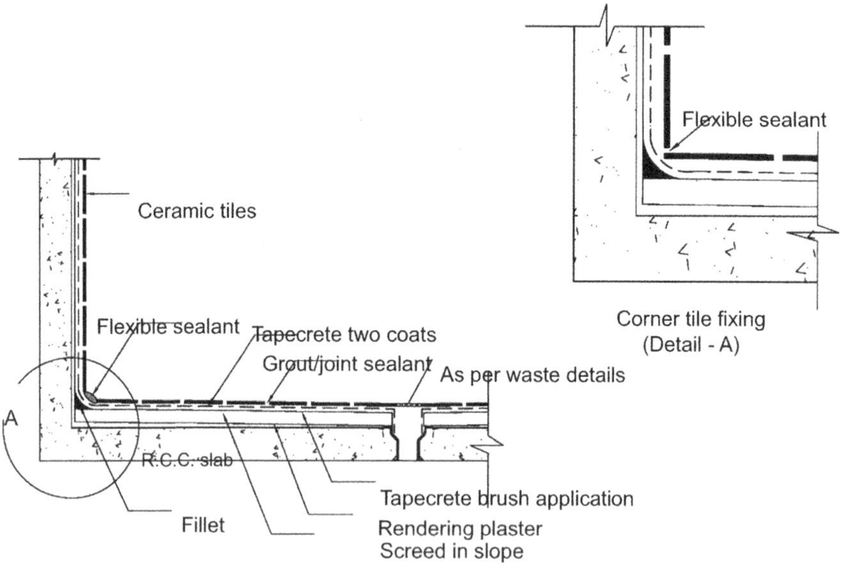

Fig. 6.7 *Waterproofing of wet areas*

Joint between flooring and WC pan should be made leak proof by using sealants like polysulphides, epoxies, etc. Similarly joints between floor and traps be made leak proof.

If clothes are to be washed on the bathroom floor, provision for that has to be provided by using a flag stone for such purpose as normal floors cannot take the wear and tear. Location of washing machine and inlet/outlet for water as well as electric connection needs attention at the planning stage.

6.7 KITCHEN DETAILS

As water tap in kitchen is kept on for washing for longer durations, pressure of water should be controlled so as not to be more than 0.17 Kg per sq.cm. Excessive pressure will result in wastage of water and inconvenience. Sink could be of stainless steel or fire clay with proper draining board. Surroundings have to be made waterproof as well. It is normal to use glazed tiles on dado up to 600 mm above working platforms as shown in the "detailed standards". (See Figs 2.8, 2.9 and 6.8)

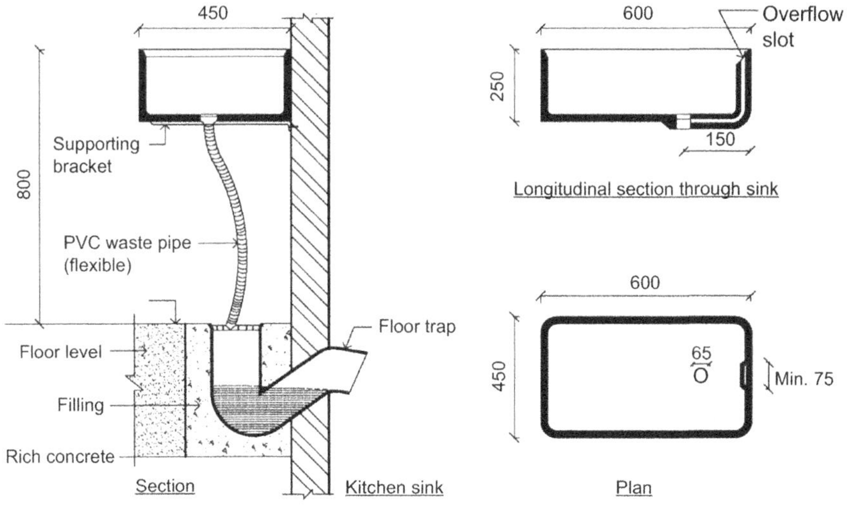

Fig. 6.8 *Fixing of a kitchen sink with the waste pipe connection to the floor trap*

Working platform should be of non-absorbent, smooth stones. Granite, marble or Kota stones are preferred. These stones should go up to wall surface and joints should be closed by tiles on vertical walls. Platform may have almost negligible slope towards outside away from wall. Height of platform should be 750–800 mm from floor. The face of the finished working platform should project 20 mm from the face of the slab (Fig. 2.7).

Proper ventilation must be provided in kitchen to avoid condensation. Use of exhaust fan is very useful. Food waste disposer/insink eretor (Electrical disposal Units) may be fitted to kitchen sink to prevent chocking of pipes (see Fig. 6.9).

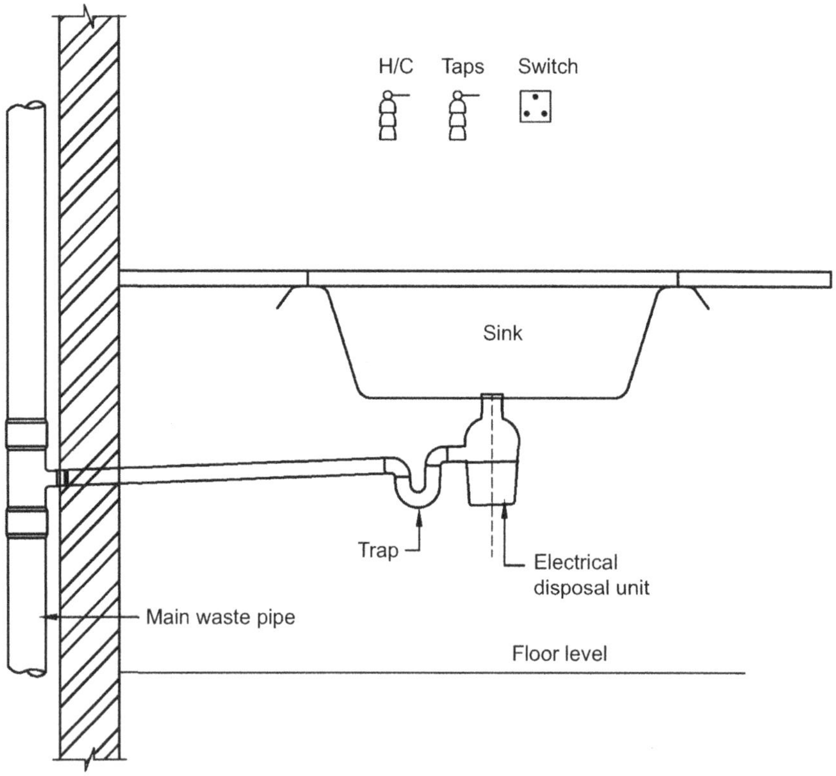

Fig. 6.9 *Electric garbage disposal unit*

CHAPTER 7

WATERPROOFING OF BUILDINGS

Waterproofing is considered to be an important part in modern construction. The word 'waterproofing' may be a misnomer as it is virtually impossible to provide absolute exclusion of water or moisture from a masonry/concrete structure. The process is basically an attempt to exclude maximum amount of water or moisture from a structure.

Waterproofing is important because it keeps a building dry, reduces internal humidity, and results in making a building more comfortable to work. It is also important to the integrity of the building itself. Typically, building waterproofing is done in several different ways to create multiple barriers for water so that it cannot enter a structure.

- Whereas waterproofing is essentially required where there is hydrostatic head of water to be resisted; damp proofing treatment is to retard the passage of water with application of suitable treatment methodology coating or integral waterproofing compounds.
- Building or structure needs waterproofing, since neither concrete nor brickwork is watertight on its own.
- Waterproofing or damp proofing of concrete or masonry is carried out in two stages. One is during construction and another is post construction. Most effective and economical waterproofing or damp proofing treatment is carried out during construction of a building whereas post construction treatment is very expensive, more time consuming and to some extent less effective.

Waterproofing may be for:

(a) Preventing hydrostatic pressure on foundations/retaining walls of basement

(b) Entry of rain water from roof or terrace garden

(c) Prevention of water seepage /leakage due to faulty pipes, fittings and fixtures in toilets and kitchens.

(d) Water ingress through walls, chhajjas and/or floors, which are not impervious.

One can get the following benefits by resorting to waterproofing and /or damp proofing:

1. Life span of the building will be increased remarkably.
2. Frequent repair or maintenance of building shall be eliminated completely or reduced to a greater extent.
3. Life of paint in interior or exterior shall be extended remarkably.
4. Aesthetic look of the building shall remain for much longer duration.
5. Minimises health hazards.

It is important to remember that waterproofing accounts for less than 1 % of the total budget foreseen for the building, but 75% of possible damages to the same building could be caused by incorrect waterproofing products and their short term durability.

7.1 WATERPROOFING MATERIALS

There are many waterproofing materials available in the market and one may get confused while choosing the right material for the right propose. However, for simplicity, we can categorise these materials as per their time of usage. First, some materials are used for waterproofing treatment as a preventive measures, and secondly, some materials are used when seepage, leakage, or dampness are noticed, i.e., for corrective measures. Agencies are available to carry out applications that suit the specific requirements. Some of the materials are discussed below but a detailed list of waterproofing materials as available in the market with names of their manufacturers and suggested applications is given in Annexure - B.

Integral Cement Water proofers: These products are incorporated in concrete or plaster during mixing with water for lowering water permeability. In normal condition, they may provide 25–90% water tightness of the structures. There are several types of integral cement water proofers available in the market. But most common varieties are:

Water reducing admixtures: These admixtures reduce the w/c ratio without altering the workability of a particular concrete mix resulting less permeable, denser concrete/masonry structures.

Very Fine Particulate Materials: Very fine particulate materials are of real benefit if the concrete mix is low in cement and in deficient in fines. However, in rich concrete mixes these materials have no advantage.

Hydrophobic/Water Repellent Agents: Materials in this group reduce the passage of water through dry concrete, which would normally occur as a result of capillary action and not as a result of an external pressure of water. In principle it is thought that all these materials impart a water repellent property to the concrete surface as well as lining and, in some cases, blocking the pores.

Air-entraining Agents (AEA): Air-entraining agents act in a similar manner to water reducing agents by imparting improved workability to the mix and thus allowing less water to be used. The micro air bubbles entrained into the cement mass shall block the capillary pores. But care shall be taken to minimise the use of over dose. Over dosing of AEA shall lead considerable amount of strength loss of concrete.

Crystalline Waterproofing Products: In this system, water bearing capillaries are blocked with insoluble crystals. Saturated concrete is given application of a solution of low chemical density crystalline waterproofing product.

Epoxy Based Waterproofing Coatings: Help in sealing pores and fine cracks. Can resist positive water pressure and resists chemical aggression.

Acrylic Polymer Based Waterproofing Compounds: Provide flexible, elastomeric layers that have good bonding with various materials and do not develop cracks.

Flexible Membranes: Produced by many manufacturers and have 82 varied applications.

APP Membranes: Improved membranes with non-woven polyester reinforcement applied gainfully in many locations.

Liquid Applied Membranes: These can be used without laps or welding and provide continuous application, very useful on roofs.

Polyurethane Water Activated Foams: These are very useful in stopping water ingress instantly and can be used in cracks or to create water proof membranes.

7.2 CHOICE OF MATERIAL

Any material that is selected for use on building for waterproofing must be checked to see that it has the following properties:

(a) Impermeable to water
(b) Breathable and seamless
(c) Good abrasion and resistance
(d) Excellent bonding properties
(e) UV resistant
(f) Non-flammable
(g) Good gap and crack bridging capacity
(h) Can be used in moist conditions
(i) Ease of application

(j) Life and durability

(k) Aesthetic considerations if used externally.

7.3 WATERPROOFING METHODS

Waterproofing methods can be broadly grouped into:

(a) Traditional Methods such as mud phuska, lime terracing, tarfelt treatment, integral waterproofing and brick coba treatment, etc.

(b) Cement/lime based treatments: Coating the surface with cement lime mortar is a time proven and economical method with good insulation properties. But it is non-flexible, and also increases the load of the structure.

(c) Mineral slurry with polymer component: It is an easy method to apply. It retains the breathing capacity of concrete but with moderate flexibility.

(d) Epoxy and polyurethane coating is highly abrasion resistant and resistant to UV radiation and does not add weight to the structure. But this has limited pot-life, not very flexible and stops breathing capacity of concrete.

(e) Elastomeric membrane forming products: It forms seamless membrane, highly flexible, UV resistant, retains breathing capacity of concrete with indirect insulation, but with low abrasion value.

(f) Silicon based impregnators as water repellent: easy to apply and economical, but do not have crack bridging capacity, and cannot withstand pressure.

(g) Bituminous based products and modified bitumen: are very economical, flexible, with good crack bridging capacity. But these products soften under heat and become brittle when cold. Life is dependent upon solvent evaporation. Other limitation is unpleasant black colour of such compounds.

(h) Improved waterproofing treatments involving use of modified membranes (APP), waterproofing chemicals. These methods can be again grouped into the following categories:

- APP membranes of various thicknesses and make.

  Usually applied over primer treated surface and torch applied.
- Cementitious slurries, which can be single component cementitious coatings or two part acrylic modified waterproofing coatings used in conjunction with cement with or without fibreglass fabric.

(i) Crystalline waterproofing system: In this system water bearing capillaries are blocked with insoluble crystals.

(j) Liquid applied membranes: Brush or spray applied to give seamless membrane.

(k) PVC membranes

(l) Rapid setting mortars used for instant plugging of seepages/ leakages.

(m) Sealants and water repellents to provide durable barrier to water penetration on porous surfaces.

In modern construction technology developments, a single product or technique is not usually enough, involvement of various bodies and techniques in co-ordination is essential for making the structure waterproof. Structures should have sufficient and efficient various control joints like expansion joints, contraction joints, etc., if proper control joints are not provided in large slabs, no waterproofing system will be successful.

7.4 COMMONLY USED METHODS

The most common waterproofing methods adopted are briefly discussed below:

7.4.1 Brickbat Coba Method

This system involves laying clay bricks with light weight lime mortar on the roof and spreading it for easy draining away of rain water. This system is popular because of the waterproofing and its weather proofing capabilities.

- In extreme weather variations its use has limited advantage.
- It also adds weight to the structure and once water starts entering, the porous clay brick pieces absorbs large quantity of water, resulting continuous leakage of roof.
- The method is described in detail while discussing waterproofing of terraces.

7.4.2 Waterproofing with Cementitious Slurries

Cementitious waterproofing membranes have been used successfully to protect wide range of buildings and structural components exposed to either periodic or long-term wetting, low hydrostatic pressure, or in combination with appropriate engineering, even high hydrostatic pressure. Cementitious membranes are used for waterproofing wet rooms and water tanks and, due to their excellent weathering resistance, also for exterior surface protection.

The advantages of cement-based waterproofing membranes are their excellent resistance to water, even if exposed permanently, their outstanding resistance to long-term weathering, good scratch resistance, good load-bearing capacity and much higher water vapour permeability compared to most other systems (consequently no danger of blistering when water vapour permeates through the waterproofing membrane).

Cement-based waterproofing slurries are easy to use, non-toxic, provide a monolithic, fully bound, joint-free surface and can easily be applied to substrates with complex surface shapes.

As polymeric binder, dispersible polymer powders are thermoplastic, plasticiser-free polymers. When water is added, these spray-dried dispersions "redisperse," while retaining all the properties and functions typical of liquid polymer dispersion. As the mortar sets, a polymer film is produced that acts as an organic binder. This greatly improves the mortar's adhesion to a wide range of substrates and increases the system's flexibility.

Today, several different systems of cementitious waterproofing membranes or slurries are available.

Standard or Rigid Mineral Waterproofing Slurries

Standard, rigid mineral waterproofing slurries are polymer-modified, pre-packed, dry mix mortars, which are gauged with water before being applied as slurry by brush, roller or airless spraying, or, if less gauging water is used, by trowel. These can only be used for where there is no risk of crack formation, movement or dimensional change (e.g., shrinkage). Such polymer-modified cementitious waterproofing membranes can withstand water pressure, not only from the positive side, but also, to a limited extent, due to their excellent adhesion and cohesion, from the negative side.

Flexible Cementitious Waterproofing Membranes (Two-component Systems)

Flexible waterproofing membranes are capable of bridging over small cracks in the substrate and can be applied to substrates expected to be subject to shrinkage, vibration, movement, etc. Due to their high polymer content, these coatings have a low coefficient of diffusion and are resistant to chemicals.

One-component Flexible Cementitious Slurries

A two-component system requires good knowledge, experience and education of the workers concerning the appropriate dosage of the liquid component. To overcome this, one component slurries are being used.

Wacker Polymers is a leading producer of state-of-the-art binders and polymer additives in the form of dispersible polymer powders and dispersions, polyvinyl acetates, surface coating resins, polyvinyl butyrals and polyvinyl alcohol solutions.

7.4.3 Waterproofing by Surface Applied Membranes

Surface-applied membrane type waterproofing system is currently accepted worldwide. The recent progress in polymer technology with the development of polymer-modified bitumen like Atactic Polypropylene (APP) modified bitumen improves the physical property of bitumen. This modified bitumen coating has overcome to a very good extent, the drawback of conventional bitumen membrane of its undesirable temperature related variations like to become brittle at freezing temperatures and soft at high temperatures.

Application of APP Membrane for Base Slab

Base slabs are constructed over levelling concrete which is applied with waterproofing membrane. External walls and roofs are covered with waterproofing membrane, which is bonded with parent concrete, and these membranes are further protected with concrete blocks protection layers coverage for any possible tearing or puncturing from external sources. Besides, there are provided construction joints and water-stops and other drainage systems as per the design requirements.

Vertical Application of APP Membrane

On sufficiently hardened, surface-leaned, blinding concrete a bituminous primer is applied. A waterproofing membrane of 4 mm thickness of homogeneous thermoplastic blend of Atatic Polypropylene (APP) distilled between with a reinforcement of 160 gm per square meter of non-woven polyester, and top surface is covered with mineral protection with silica and the bottom surface is covered with a flammable polyethylene film, is then thermo fused by flame torching of inside face of membrane, which is embossed and protected by a seven micron thick polyethylene film. On this membrane-applied surface, a high abrasion resistant polymeric mortar is laid up to 3 mm thickness. At least three days curing is provided to this mortar layer before laying reinforcement.

For walls, on a clean and dry surface a primer is applied. When it is touch dry, the similar APP membrane is thermo fused by flame torching. To protect this membrane system from any further damage during other external activities, a 75 mm thick concrete block work is provided. For roof structures the similar methods followed for membrane application and protection layer of concrete is provided.

7.4.4 Waterproofing with Crystalline Technology

Concrete is often susceptible to damage and deterioration from water and chemical penetration. These deleterious effects of concrete can be avoided through the use of crystalline waterproofing technology, which effectively improves the durability and life span of concrete structures, thereby reducing long-term maintenance costs.

Despite its apparent density, concrete remains a porous and permeable material that can leak and deteriorate rapidly when in contact with water or the intrusion of aggressive chemicals, such as carbon dioxide, carbon monoxide, chlorides, sulphates or other substances.

Crystalline waterproofing technology improves the waterproofing and durability of concrete by filling and plugging pores, capillaries, micro-cracks and other voids with a non-soluble, highly resistant crystalline formation.

When concrete is saturated with water prior to applying crystalline waterproofing, a solution of low chemical density is also applied. When crystalline waterproofing is applied to the concrete, a solution of high chemical density is created at the surface, triggering the process of chemical diffusion. The crystalline waterproofing chemicals migrates through the water (the solution of low density) until the two solutions equalise .

Chemical diffusion will take these chemicals about 12 inches into the concrete. If water has only soaked two inches into the surface, then the crystalline chemicals will only travel two inches and stop; but they still have the potential to travel 10 inches further if water re-enters the concrete at some point in the future and reactivates the chemicals.

Instead of reducing the porosity of concrete, like water reducers, plasticisers and super plasticisers, the crystalline formation engages the material filling and plugs the voids in concrete to become an integral and permanent part of the structure.

Negative Side Waterproofing

Existing basements that are subject to water seepage or vapour transmission through foundation walls and floors can be treated by the application of crystalline waterproofing and protection on the negative side or the inside of the structure. Surface coatings will blister and peel when moisture seeping through the concrete dissolves soluble minerals and deposits them on the surface under the coating, in the form of efflorescence—a white powdery substance that forms on the wall surface. Because crystalline waterproofing penetrates into the concrete plugging the pores beneath the surface, it does not depend on surface adhesion, and will not blister and peel off unlike surface barriers.

7.5 STEPS BEFORE APPLICATION

Desirable:

In all waterproofing applications, certain steps are common and have to be ensured for success of any treatment. These have been listed here:

Step 1: Ensure that the structural slab or member to be treated for waterproofing is structurally sound and has no voids. Especially in slabs, it is suggested that in addition to using good quality concrete with controlled water cement ratio, injection grouting should be done to ensure impermeability of the slab.

Step 2: Provide slope of at least 1:100 in structural member itself if possible or else with screed/foam concrete.

Slope to be towards the shaft.

Step 3: In depressed slabs place a 25 mm diameter pipe for water to ooze out into shaft.

Step 4: Provide fillets/haunches on the entire vertical/horizontal joints with a polymer admixed mortar over polymer slurry.

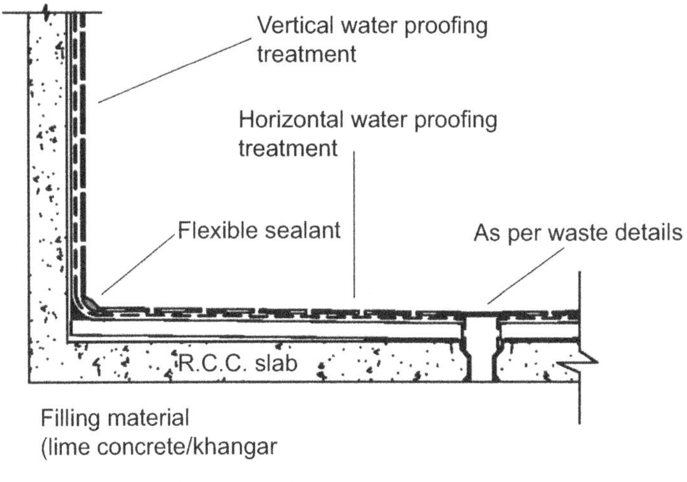

Fig. 7.1 *Position of horizontal and verticalwaterproofing treatment in sunken portion of WC/kitchen and the like*

Step 5: The surface to be treated shall be cleaned of all dirt, grease, foreign matters and loose materials. To achieve this one should resort to mechanical cleaning like scarifying, grinding, water or sand blasting, or chemical cleaning, or any other approved method. Wash the surface to clean thereafter.

Step 6: Surface has to be made reasonably smooth so as not to impede the application.

Step 7: If there are any cracks or damaged areas, these have to be repaired/rectified before waterproofing treatment.

Step 8: Paint all pipes with polymer compound being used for waterproofing ensuring that it is taken two inches inside the pipes. This will avoid corrosion and deterioration of pipes.

Step 9: Lay minimum 50 mm thick concrete screed for laying the floor using 10 mm aggregate.

Step 10: Use waterproof tile grout for all tile joints.

Step 11: Ensure proper slope (1:50) in all floors of toilets, kitchens, balconies etc. Roofs can have a slope of 1:100.

Step 12: Ensure installation of all traps, outlets before waterproofing treatment. All floor traps, etc., have to be properly grouted with rich mortar mixed with waterproofing admixture/ waterproof sealants.

Step 13: Seal with butyl sealant all joints of the fittings to the floor as also the joints of the traps to the floor.

Step 14: Seal the joints of the bathtub and the wall with a cartridge silicone sealant.

Step 15: Proper bedding and hunching of drainage and sewerage pipes must be ensured. The surface on which these are laid must be stable and should not be liable to flooding/sinking.

Avoid the following:

(i) Use of tiles of the water absorbing type such a slate, etc.

(ii) The practice of encasing the C.I. pipes in concrete, as the same offers no advantages whatsoever and creates weak points.

(iii) Filling up the sunken portions with cinder.

7.6 PREVENTIVE OR RECTIFICATION MEASURES

Some materials are used for waterproofing treatment as preventive measures at the construction stage and there are other materials that are used when seepage, leakage, or dampness has been detected, and are required for corrective measures. Preventive measures are adopted at construction stage and will be most effective both in cost and operation.

7.7 WATERPROOFING OF BASEMENTS

The structural members have to be designed for the dead and live load as well as for the Hydrostatic pressure keeping the worst conditions in view. Overburden pressure of fills in Basements should not be considered in designs unless it is ensured that it will be in place till waterproofing operations are completed and tested fully and also during service life of the building (Fig. 7.2).

Following methods can be adopted for waterproofing of basements:

7.7.1 Integral Cement Based Waterproofing (with Rough Kota Stone)

(A) For Horizontal Surface

Nomenclature of item

Providing and laying integral cement based treatment for waterproofing on horizontal surface at all depth below ground level for underground structures as directed by the engineer-in-charge and consisting of:

(i) First layer of 25 mm thick approved and specified rough Kota stone slab over a 25 mm thick base of cement mortar 1:3 (1 cement : 3 coarse sand) mixed with waterproofing compound conforming to IS:2645 in the recommended proportion over the levelling course (course to be paid separately). Joints sealed and grouted with cement slurry mixed with waterproofing compound.

(ii) Second layer of 25 mm thick cement mortar 1:3 (1 cement: 3 coarse sand) mixed with waterproofing compound in recommended proportions.

(iii) Finishing top with stone aggregate of 10–12 mm nominal size spreading @ 8 cudm/sq m thoroughly embedded in the second layer.

Methodology: Preparation of Surface

(i) The surface of levelling course should be roughened properly when the concrete is still green. In case the surface is not made rough before the concrete is set, the work of waterproofing should be executed only after providing proper key for the base layer of cement mortar 1:3 as it is essential that the waterproofing treatment adheres to the lean concrete/ levelling course surface firmly.

(ii) Before laying the base course of cement mortar 1:3, the lean concrete surface shall be cleaned neatly with water.

(iii) Cement slurry shall be prepared by using 2.2 kg of blended cement per sq m. area. Cement slurry shall be applied only on the area of the concrete surface that can be covered with the cement mortar (1:3) base course within half an hour. The cement slurry should cover every spot of the surface and no place shall remain uncovered.

(iv) Just after the application of cement slurry on the surface, Cement mortar 1:3 (1 blended cement: 3 coarse sand) shall be prepared with cement/water duly blended mixed with waterproofing compound conforming to IS:2645 in recommended proportion should be used for laying the base course. Base Course should be laid to a perfect level with wooden/aluminium straight edge of at least 2 m. long. The top surface of cement mortar should be finished neatly and later scratched when green with a suitable instrument before the base course dries and gets hard that is just before the base course takes up initial set.

(v) Slabs shall then be set to the level of the string. When the 25 mm thick base course is just getting set the cement slurry prepared, as detailed above, should be spread over the base course up to the area that shall be covered with just two to three stone slabs. The cement slurry shall be spread in such a way that the area of base course to be covered immediately shall be covered with slurry without any gap or dry spots. Immediately on applying cement slurry on the base course first layer of 25 mm thick is approved and specified rough stone slab of large size, i.e., 550 mm x 550 mm or 550 mm x 850 mm so as to minimise the number of joints shall be laid over the base course and pressed gently so that the air gap can be removed. (General requirement of Kota stone shall be as laid down in CPWD Specifications for Kota stone flooring.) The slurry applied on the surface which gets spread when the stone slab is pressed shall get accumulated in the joints of adjacent stone slabs and if any gap still remains between the stone slabs the same should also be filled with additional quantity of cement slurry. For laying the stone slabs in perfect level, two stone slabs at adjacent concerns/ends shall be fixed firmly to the required level and a string stretched over the two slabs, the intermediate.

(vi) After filling all the joints of the rough Kota stone slabs with cement slurry, the surface of stone slabs shall be cleaned and lightly watered. The area of stone slab shall be laid with second layer of 25 mm thick cement mortar 1:3 (1 cement: 3 coarse sand) mixed with waterproofing compound in recommended proportions.

(vii) Before the mortar layer takes the initial set, top shall be finished with stone aggregate of 10–12 mm nominal size uniformly spread @ 8 cudm / sq m and lightly pressed into the finished surface thoroughly embedded in the second layer. The aggregates shall not be embedded totally inside the mortar, and shall be visible on the top surface.

(viii) In cases where slope is to be provided for the waterproofing layer, grading with additional cement concrete/cement mortar shall be provided, and then the waterproofing layer shall be laid on the graded surface. Extra payment shall, however, be made for the grading course.

(ix) Arrangements should be made to lay RCC slabs as quickly as possible after the last operation of Kota stone waterproofing is over. RCC slabs must be checked for their soundness by carrying out grouting operation through grouting nozzles left at specified

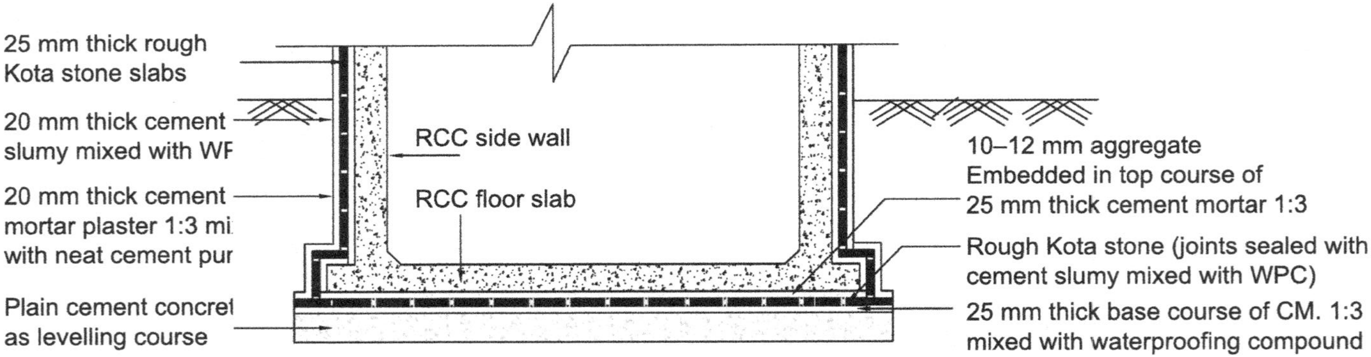

Fig. 7.2 *Waterproofing of basements with rough Kota stone*

spacing in a grid for grouting purposes. Waterproofing treatment shall be cured for 14 days or till the RCC slab is cast, whichever is earlier.

(B) For Vertical Surface

Nomenclature of item:

Providing and laying integral cement based treatment for waterproofing on vertical surface by fixing specified stone slab of 25 mm thick of size not exceeding 550 x 400 with cement slurry mixed with waterproofing compound conforming to IS:2645 in recommended proportions with a gap of 20 mm (minimum) between stone slabs and the receiving surfaces and filling the gaps with neat cement slurry mixed with waterproofing compound and finishing the exterior of stone slab with cement mortar 1:3 (1 cement : 3 coarse sand) 20 mm thick with neat cement punning mixed with waterproofing compound in recommended proportion complete at all levels.

Methodology:

(i) Preparing the surface

The surface of the structure to be treated shall be roughed either by raking of joints in case of brick/ stone masonry or by hacking the cement concrete surface with a specifically made hacking tool just after removing shuttering.

(ii) Treatment of construction joints

Special attention is required to be paid to the construction joints in the concrete walls. If the joint is not properly finished then a 'v' groove should be cut to a depth of 25 mm and filled with joint sealant. Further coats of cementitious crystal waterproofing application with material like Krystol T1 at the rate of 0.8 kg/sq m over 200 mm width of the joint line covering areas on both sides of joint are provided as per manufacturer's specification before taking up the Kota stone waterproofing treatment. The method is described in detail in Crystalline applications.

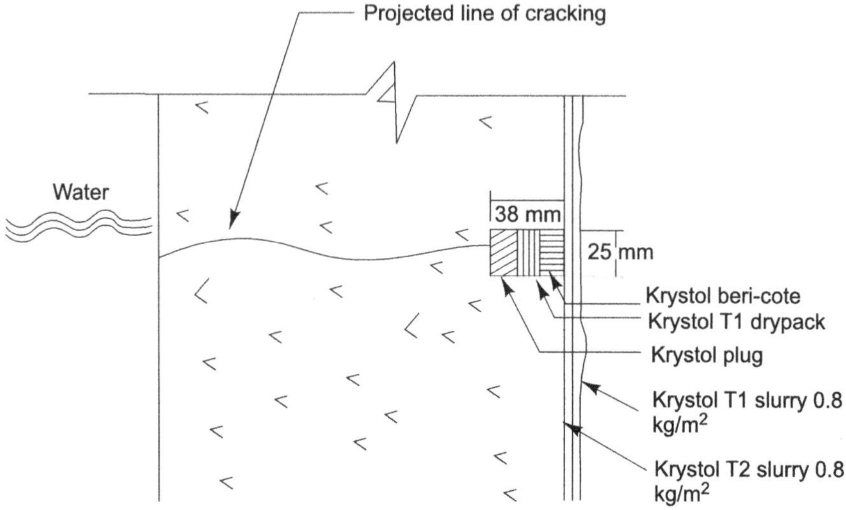

Fig. 7.3 *Crack repair*

Where hydrostatic pressure is high, it is recommended that Swelling Rubber Sealings manufactured by WEBAC 20 × 6 mm rectangular profile can be placed over the already cast concrete just inside the outer surface before laying fresh concrete on top. This has a property to swell when in contact with water and seals the joint completely.

(iii) If rough Kota stone are not sufficiently rough to remain in vertical position held by cement slurry. Therefore, the grip for the stone slab has to be increased and this can be done by planting 12–15 mm nominal size stone aggregate fixed with araldite on surface of each sand stone slab.

(iv) Process

- The surface preparation, slurry preparation and making the Kota stone rough shall be as for horizontal surfaces. Kota stones shall be erected vertically (longer side to be kept horizontal) keeping a gap of 20 mm between RCC wall and the Kota stone face. To ensure proper gap and vertical placement of stone, stiff cement mortar of mix 1:3 (1 cement and 3 coarse sand) shall be applied to centre of stone in an area of 50 ↔ 50 mm.

The gap shall be filled in with Cement Slurry properly blended with waterproofing compound, to a height of three fourth of the height of stone after completion of one layer. Next stone on top of the first shall be fixed same way on the next day to allow lower layer to set. As work proceeds, the gap will be continued to be filled to a level lower than the top of stone slab by 1/4th of the stone height.

- After completing the vertical surface, the surface of stone slabs shall be cleaned and lightly watered. The area of stone slab shall be laid with a layer of 20 mm thick cement mortar 1:3 (1 cement: 3 coarse sand) mixed with waterproofing compound in recommended proportions. Neat cement punning mixed with water proofing compound in recommended proportion shall be applied after completion of mortar application on green surface.
- Curing shall be carried out for 14 days before backfilling is done. Refer Fig. 7.2.

C. Limitations

Though, this type of stone is cheap and good water-resistant, there are many joints between stone slabs from where water may percolate to the main concrete and need to be filled in properly. Another major disadvantage is it is very difficult to rectify the post construction defects.

7.7.2 Alternative Methods

Alternative methods include application of products manufactured by different firms such as Liquid applied membrane, APP, which is torch applied and crystalline products. These are to be used in specific cases in consultation with manufacturers and site conditions. Care is to be taken to see that surfaces are dry for applications of membranes as these are sensitive to moisture and will not stick to damp surfaces. If dry surface cannot be achieved then crystalline method is recommended.

(a) APP membrane application as per CPWD specifications.

(b) Liquid membrane application as per manufacturer's specifications. It provides seamless without overlaps. These membranes are available from M/s KEMPEROL (Kemerol V 210), BASF, ZUPA Concrete Material Science (Derma Rubber) Mumbai.

(c) Use of Glass fibre reinforced waterproofing membranes or PVC/ TPO preformed membranes manufactured by M/s BASF may be used with protective layer.

7.8 FOR TERRACE GARDENS, SWIMMING POOLS AND EXTENDED BASEMENTS

As terrace gardens and extended basements are likely to have sustained water accumulation, it is suggested that in addition to taking all the preliminary steps suggested above such as good concrete, grouting and slopes, extra care is given to the waterproofing treatment. Following are suggested:

(a) Apply Crystalline waterproofing treatment over prepared surface of slab. Method is described below

(b) Provide grading concrete to ensure slopes in case of extended basements.

(c) Install drainage pipes, etc.

(d) Give protection by suggested architectural finish or lay the roof garden as the case may be.

(e) In case very heavy water ingress is anticipated then an additional layer of waterproofing treatment by rough Kota stone or of APP or Liquid membrane may be provided.

(f) The treatment provided should be roots resistant so that plant roots do not damage the waterproofing treatment.

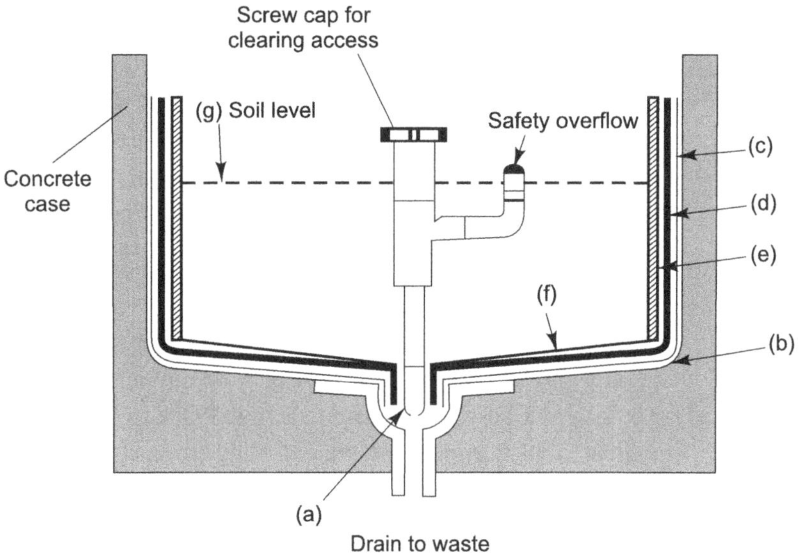

Fig. 7.4 *Planter box*

(a) Grate to stop roots going down drain waste.
(b) Cracks to be treated first with silicone or sodaflex 603.
(c) Prime with suitable primer.
(d) Apply the membrane over the entire surface.
(e) Fibre cement protection board to stop root damage to the membrane.
(f) Drainage cell - recommended that this be wrapped in geo-filter fabric, to protect the membrane from root damage, and aid in drainage of moisture to the outlet.
(g) Soil level should be 100 mm below the membrane.

7.9 CONCRETE WATERPROOFING WITH CRYSTALLINE TECHNOLOGY

A. Crystalline waterproofing system: In this system water bearing capillaries are blocked with insoluble crystals; the saturated surface is applied with one or two coatings with crystalline waterproofing slurry.

(a) Materials

Various proprietary materials are now available. One of the manufacturers is M/s Kryton Buildmat Co. Pvt. Ltd, with product range consisting of Krystol T1 and T2, Krystol Bericote and KIM (Krystol Internal Membrane). For stopping of water Krystol Plug is recommended. Similar product is also marketed by M/s Choksey Chemicals P. Ltd by the trade name of Drizoro.

Applications include waterproofing of Rafts, retaining walls, terraces and we areas. It is especially used when the seepage is required to be stopped and this material can be used from negative side as well. Here we are suggesting use for terrace gardens.

(b) Nomenclature of Item

Providing and applying of cementitious crystalline based waterproofing for concrete rafts, walls, water tanks, slabs, concrete floors, etc., with Krystol T1/T2 system. The application is to be done from +ve (positive) side on a wet open pore concrete surface with brush @0.8 kg of Krystol T1 and 0.8 kg of Krystol T2 as per manufacturer specification.

Construction joints treatment - A U - shaped groove will be made of 10 mm wide and 15–20 mm deep. This will be filled up with Krystol T1 putty consistency 5:1 (5 parts of T1 to 1 part of water) and Krystol Baricote in 3.5:1 (3.5 part of Baricote to 1 part of water).

(c) Procedure

Surface Preparation and Krystol T1/T2 Application

- Concrete surfaces will be cleaned and made free of contaminants and laitance.
- Concrete must be sound.

- Concrete will be cleaned to have an open pore surface to allow penetration of Krystol. This may require mechanical preparation such as grinding water.
- blasting, sandblasting, or hacking. Surfaces to be treated will be pre-soaked with clean water to a saturated surface dry (SSD) condition. Do not leave any standing water.
- Mix Krystol T1 to a slurry consistency (5 parts powder to 2 part clean water). (Mix only enough material that can be placed in 15–20 minutes)
- Apply slurry to the concrete surface with a brush in a circular scrubbing motion so as to achieve maximum adhesion and penetration. Apply the slurry to the entire concrete slab. Krystol T1 with a spread @ rate of 0.8 kg/m2.
- Mix Krystol T2 to a slurry consistency (5 parts of Krystol T2 to 2 parts of clean water) mix enough which can be applied in 15–20 minutes.
- Apply Krystol T2 slurry on the concrete surface where Krystol T1 treatment has been completed, with brush in a circular scrubbing motion so as to achieve maximum adhesion and penetration. The Krystol T2 spread @ 0.8 kg /m2. This may be applied after 2–3 hours when Krystol T1 coating has set hard.
- Protect the area from rapidly drying out due to heat, damage by rain, excessive wind and freezing temperatures for 48 hours.
- Testing with pounding to be done after 7 days.

For construction joint treatment a U- shaped groove will be made of 10 mm wide and 15–20 mm deep. This will be filled up with Krystol T1 putty consistency 5:1 (5 parts of T1 to 1 part of water), and Krystol Baricote in 3.5:1(3.5 part of Baricote to 1 part of water) consistency as per manufacturer's specification. First half of the joint is filled by Krystol T1 and balance portion by Krystol Baricote.

7.10 METHODOLOGY FOR PREVENTION OF FLOWING WATER

(a) Nomenclature of Item:

Waterproofing treatment with KRYSTOL T1 system providing and applying cementitious crystalling waterproofing treatment to the concrete wall, etc., to make the structure impervious to the flow of liquid through it. Coating with Krystol T1 at the rate of 1.0 kg/sq m over the surfaces after mixing uniformly in ratio of 5:2 (5 parts of krystol T1:2 parts clean water). Construction joints shall be treated with krystol T1 system. Material shall be mixed and application to be done as per manufacturer specifications.

(b) Suggested Precautions:

(i) The treatment should be protected to avoid and damage to surface and also provide first line of defence.

(ii) Before any treatment all cracks have to be repaired.

(c) Methodology:

Following procedure is recommended by the manufacturers of Krystol:

Step 1: Prepare the Crack or Joint

(i) Using a sharp 25 mm wide chisel, chip a 25 mm wide chase along the entire length of all cracks to a minimum depth of 40 mm. The shape of the chase is critical to success. The chase must be square shaped and deeper than its width. If the concrete breaks apart near the surface, chisel deeper to obtain the required 25 mm by 40 mm size and shape.

When chiselling, the chisel should not be placed inside the chase. Instead, the chisel should be placed on the concrete surface over the crack about one inch ahead of the chase and directed to ensure that the chisel pressure is towards the chase so that the piece being removed falls into the chase. Chisel to the full depth of 40 mm before moving on. This method will result in a chase that is the proper shape.

(ii) Wash chase with water so that it is clean. Use a vacuum if necessary to remove dust, debris or water.

(iii) Be sure to repair the full length of the crack.

IMPORTANT: Steps 2, 3 and 4 must all be completed on the same day. Do not begin step 2 unless you can complete step 4.

Step 2: Stop any Flowing or Seeping Water

(Note: Step 2 can be skipped if the crack is dry at the time of the repair)

(i) Mix plug (Krystol Plug™ For every 30 m of crack approx. one 25 kg pail of each product is required) to a putty consistency (4 parts powder to 1 part water). Mix only enough material as can be placed in 1 minute.

(ii) Using a gloved hand, press the putty into the chase stopping the water in that spot. Hold without moving until set. Do not move or work the plug after pressing it into place or it will come apart.

(iii) Repeat this procedure, working from one end of the crack to the other until all flowing or seeping water has been completely stopped.

(Areas of very high water flow should be left to last. Insert a rubber hose at the worst location and install plug around it. Removing the hose will leave a deep narrow hole that is much easier to Plug with a single ball of material.)

IMPORTANT: When finished, the plug must not fill more than one-third of the depth of the chase. If the Plug fills more than one-third, use a chisel to remove the excess Plug immediately so that at least one inch of space remains in the chase. Refer to Fig. 7.3.

Step 3: Install Krystol T1®

(i) Ensure that the concrete is saturated with water, but that no water remains on the surface.

IMPORTANT NOTE: A saturated surface-dry (SSD) condition is extremely important to success. The concrete must be completely saturated with water to allow the Krystol® chemicals to penetrate deeply and react. The outer surface, however, must be only slightly damp, so as not to dilute and weaken the bond of the Krystol application.

(ii) Mix Krystol T1® to a very dry consistency (approximately 5 parts powder to 1 part clean water). Use as little water as possible.

(iii) Pack the Krystol T1® tightly into the chase to a maximum thickness of 13 mm.

(iv) The chase should now be maximum two-thirds full (or one-third full if step 2 was skipped). It is essential that enough space remains to install at least 13 mm of Krystol Bari-Cote™.

Step 4: Finish Flush with Krystol Bari-Cote™

(i) Mix Krystol Bari-Cote™ to a stiff putty consistency (approx. 4 parts powder to 1 part clean water). Mix only as much as can be placed in 15 minutes.

(ii) Ensure that the concrete is in saturated surface-dry (SSD) condition.

(iii) Use a trowel to fill the rest of the chase with Bari-Cote flush to the surface (outer third or outer two-thirds if step 2 was skipped).

Step 5: Apply Krystol T1® Slurry Coat

(i) Mix Krystol T1® to a slurry consistency (5 parts powder to 2 parts clean water). Mix only as much as can be placed in 30 minutes.

(ii) Ensure that the concrete is in a saturated surface-dry SSD condition.

(iii) With a concrete brush, use an aggressive, circular scrubbing motion to apply the T1 slurry to the repair area, extending at least 6 inches to either side of the repair. Apply at 0.8 kg/sq m.

(It is highly recommended that the entire wall, floor and/or ceiling be treated with the Krystol T1® slurry coat followed by a slurry coat of Krystol T2®.)

(iv) Krystol growth and migration only occurs in the presence of water. For this reason Krystol T1/T2 must be "wet (moist) cured" for at least 3 days (continued curing for several days or even weeks will be beneficial in most cases). During the curing period, treatment area must be protected from frost, rain and traffic at least 24 hours.

The vertical and horizontal surfaces need to be protected if these are likely to be exposed to abrasive forces as in case of parking, etc., Manufacturers do not recommend protection otherwise.

7.11 NEGATIVE SIDE WATERPROOFING

Existing basements that are subject to water seepage or vapour transmission through foundation walls and floors, can be treated by the application of crystalline waterproofing and protection on the negative side, or the inside of the structure.

7.12 WATERPROOFING OF TERRACES

Before taking up the waterproofing work, the construction of parapet walls, including finishing should be completed. Similarly, the ancillary items like haunches, khurras, grooves to tack the fibre cloth layer, fixing up of all down take pipes, water pipes and electric conduits, etc., should be completed and no such work should be allowed on the area to be treated during the progress of waterproofing treatment or even later.

7.12.1 Integral Cement Based Waterproofing Treatment with Brick Bat Coba

(A) Nomenclature

Providing and laying integral cement based waterproofing treatment including preparation of surface as required for treatment of roofs, balconies, terraces, etc., consisting of following operations: (a) Applying

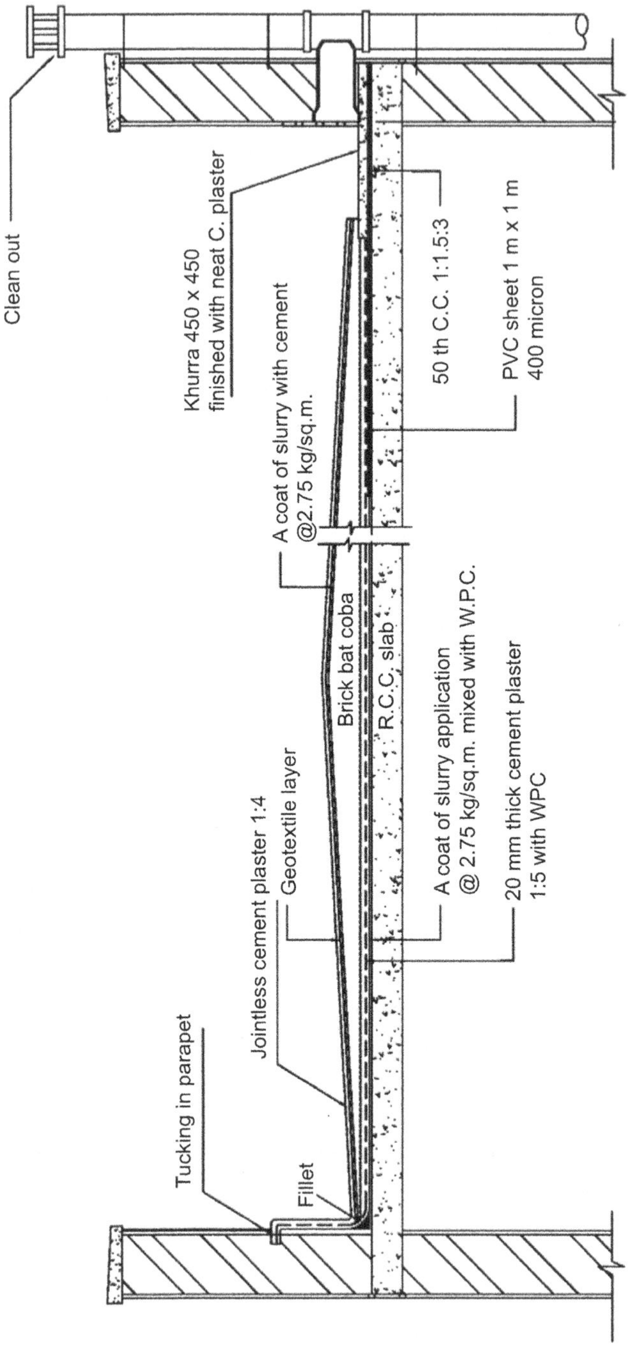

Fig. 7.5 *Brick bat coba treatment*

a slurry coat of neat cement using 2.75 kg/sq m of cement admixed with waterproofing compound conforming to IS. 2645 over the RCC slab including adjoining walls up to 300 mm height including cleaning the surface before treatment. (b) Laying brick bats with mortar using broken bricks/brick bats 25–115 mm size with 50% of cement mortar 1:5 (1 cement : 5 coarse sand) admixed with waterproofing compound conforming to IS : 2645 over 20 mm thick layer of cement mortar of mix 1:5 (1 cement :5 coarse sand) admixed with waterproofing compound conforming to IS : 2645 to required slope and treating similarly the adjoining walls up to 300 mm height including rounding of junctions of walls and slabs (c) After two days of proper curing applying a second coat of cement slurry using 2.75 kg/sq m of cement admixed with waterproofing compound conforming to IS : 2645. (d) Finishing the surface with 20 mm thick joint less cement mortar of mix 1:4 (1 cement :4 coarse sand) admixed with waterproofing compound conforming to IS : including laying glass fibre cloth of approved quality/ geotextile layer as approved by engineer in top layer of plaster and finally finishing the surface with trowel with neat cement slurry and making pattern of 300 x 300 mm square 3 mm deep. (e) The whole terrace so finished shall be flooded with water for a minimum period of two weeks for curing and for final test. All above operations to be done in order and as directed and specified by the Engineer-in-Charge with average thickness of 120 mm and minimum thickness at khurra as 65 mm.

(B) Methodology

Providing and laying integral cement based waterproofing treatment including preparation of surface as required for treatment of roofs, balconies, terraces, etc., consisting of following operations:

(a) Preparing the Surface

The surface of the slab should be roughened by scrapping when the slab concrete is still green, however, the surface need not be hacked. In case the slab is already cast and surface fairly finished, the same shall be cleaned neatly of all mortar droppings, loose materials, etc., with brooms/cloth.

(b) Providing and Laying of Slurry under Base Coat

Depending upon the area of surface that has to be covered, the required quantity of slurry should be prepared using 2.75 kg blended cement admixed with waterproofing compound conforming to IS: 2645+ water per sq m. area to be covered, taking particular care to see that only that much quantity of slurry shall be prepared that can be used within half an hour of preparation, i.e., before the initial setting time of cement.

The prepared slurry shall be applied over the dampened surface with brushes very carefully, including the joints between the floor slab and the parapet wall, holes on the surfaces, joints of pipes, masonry/concrete, etc.

The application of the slurry should continue up to a height of 300 mm on the parapet wall and also the groove as shown in Fig. 7.5.

(c) Laying Base Coat 20 mm Thick

Immediately after the application of slurry and when the application is still green, 20 mm thick cement plaster as base coat with cement mortar 1:5 (1 blended cement : 5 coarse sand) admixed with waterproofing compound conforming to IS: 2645 shall be evenly applied over the concrete surface taking particular care to see that all the corners and joints are properly packed and the application of the base coat is continued up to a height of 300 mm over the parapet wall.

(d) Laying Brick Bat Coba

Brick bat of size 25–115 mm out of well burnt bricks shall be used for the purpose of brick bat coba. The brick bats shall be properly dampened for six hours before laying.

Brick bats shall be laid to required slope/gradient over the base coat of mortar leaving 15–25 mm gap between two bats. Cement mortar 1:5 (1 blended cement: 5 coarse sand) shall be poured over the brick bats and joints filled properly in two parts—first half thickness and then the balance half. Under no circumstances dry brick bats should be laid over the base coat.

The haunches/gola at the junction of parapet wall and the roof shall be formed only with brick bat coba as shown in Fig. 7.5.

In case the brick bat coba is laid on the base coat immediately on initial set there will be no necessity of applying cement slurry over the base coat before laying the brick bat coba. However, if the brick bat coba is to be laid on the subsequent day, cement slurry prepared as described earlier shall be applied over the top surface of the base coat, then only the brick bat coba shall be laid.

(e) Application of Slurry over Brick Bat Coba

After two days of curing of brick bat coba, cement slurry admixed with waterproofing compound conforming to IS: 2645 shall be applied on the surface of brick bat coba The application of slurry shall cover the haunches/gola, and the remaining small portion of parapet wall and also inside the groove as shown in the Fig. 7.5.

(f) Laying Finishing Layer (Protective Coat)

Immediately on applying the cement slurry over the surface of the brick bat coba and when the slurry applied is still green, the fibre glass cloth of approved brand, which shall be thin, flexible, uniformly bonded mat composed of chemically resistant borosilicate glass fibre distributed in random open porous structure bonded together with a thermosetting resin, shall be spread evenly on the surface without any kink and pressed to see that no air spaces exist. The fibre glass cloth shall be taken up to a height of 300 mm on parapet walls and tucked in the groove specially prepared at that height. 20 mm thick joint less cement mortar of mix 1:4 (1 cement: 4 coarse sand) admixed with waterproofing compound conforming to IS: 2645 and approved by the engineer-in-charge, shall be applied without leaving any joints over the entire fibre glass cloth including the haunches/gola and the small portion on the parapet wall. The groove in the parapet wall over the haunches shall also be filled neatly packing the mortar firmly in the groove.

The surface of the finishing layer (protective coat) shall be neatly finished with cement slurry. The finished surface shall be allowed to dry for a while and then pattern of 300 mm x 300 mm groove, 8 mm deep shall be made over the entire surface.

(g) Curing and Testing the Treatment

The entire surface thus treated shall be flooded with water by making kiaries with weak cement mortar, for a minimum period of two weeks.

Normally Brick coba treatment is done with average thickness of 120 mm and minimum thickness at Khurra as 65 mm. Refer to Fig. 7.5.

7.12.2 Alternative Methods

Use of waterproofing treatment with atactic polypropylene polymer (APP) modified prefabricated membrane is also recommended for terraces. Details are given below.

(A) Waterproofing Treatment with Atactic Polypropylene Polymer (APP) Modified Prefabricated Membrane

Now-a-days factory made polymer modified bituminous membrane (APP - Atactic Poly Propylene) sandwiched with synthetic fibre cloth, polymer modified PVC membrane, neoprene rubber membrane, etc., are available in the market for barrier and long lasting. These membranes are available in roll form with different thickness (1.5–5 mm) and may be fully bonded to the substrate or laid loose according to the manufacturer's instructions. Elongation is 40–50%. The main drawback is very poor vapour breathability. Therefore, one should ensure that the structure should be free from moisture as much as possible inside the structure before laying the membrane.

Nomenclature of item

Providing and laying 3 mm thick APP (Atactic Polypropylene Polymer) modified prefabricated five layer 3 mm thick waterproofing membrane, black finished reinforced with non-woven polyester matt consisting of a coat of bitumen primer for bitumen membrane @ 0.40 litre/sq m. by the same membrane manufacture of density at 25°C, 0.87–0.89 kg/litre and viscosity 70–160 cps. Over the primer coat the layer of membrane shall be laid using Butane Torch and sealing all joints, etc., and preparing the surface complete. The vital physical and chemical parameters of the membrane shall be as under: Joint strength in longitudinal and transverse direction at 23°C as 650/450N/5 cm. Tear strength in longitudinal and transverse direction as 300/250 N. Softening point of membrane not less than 150°C. Cold flexibility shall be up to -2°C when tested in accordance with ASTM, D - 5147. The laying of membrane shall be done through the authorised applicator of the manufacturer of membrane.

Materials

Bitumen primer for bitumen membrane shall have density at 25°C in the range of 0.87–0.89 kg./ and viscosity of 70–160 cps primer shall be applied @ of 0.40 litre/sq m.

Atactic Polypropylene Polymer Modified Prefabricated Membrane: It is a polymeric waterproofing membrane. This shall be 3 mm thick with non-woven polyester matt.

Physical and chemical parameters of the membrane shall be as given in relevant specifications.

Preparation of Surface: The surface to be treated shall have a minimum slope of 1 in 100 or as specified. This grading shall be carried out with cement concrete or cement plaster with coarse sand, as desired, to the average thickness required and finished smooth. Such grading shall be paid for separately.

Junctions between the roof and vertical faces of parapet walls, chimneys, etc., shall be chased by running triangular fillets 7.5 x 7.5 cm size, cement concrete. At the drain mouths, the fillets shall be suitably cut back and rounded off for easy application of waterproofing treatment and easy flow of water. Cement concrete where shall be 1:1.5:3 mix (1 Cement: 1.5 Coarse sand: 3 Graded stone aggregate 20 mm. Nominal size). The provision of fillets shall be deemed to be covered by the item of waterproofing and shall not be measured or paid for separately.

While the grading of roof surface is being done, it shall be ensured that the outlet drain pipe has been fixed and mouth at the entrance have been eased and rounded off properly for easy flow of water. When any pipe passes through the roof to be treated, angular fillet shall be built around it for the waterproofing treatment to be taken over it. For carrying over and tucking in the waterproofing membranes into the parapet walls, chimneys stacks, etc., a horizontal groove 6.5 cm deep, 7.5 cm wide section with its lower edge at not less than 30 cm above the graded roof surface shall be left on the inner face of the same; during construction if possible. When such groove has not been left, the same shall be cut out neatly and the base at rear of the groove shall be finished smooth with cement plaster 1:4 (1 cement: 4 coarse sand). Such cutting of the groove and its finishing smooth shall be part of the waterproofing. Tucking in the waterproofing membrane will be required where the parapet wall exceeds 45 cm in the height from the graded surface. Where the height is 45 cm or less, no groove will be required as the waterproofing treatment will be carried over the top of the parapet wall to its full thickness.

For grading of roof use of foam concrete with styropore beads @ 16–17 kg per cum of concrete can be made to have light weight concrete for providing slopes on roofs. This also gives thermal insulation.

To ensure good adhesion between the surface and waterproofing treatment, suitable method to dry the surface shall be adopted. All hair line and other cracks in the surface should be filled with approved sealant.

Laying: Bitumen primer @ 0.40 litre/sq m shall be applied to the prepared roof, drain and all other surfaces where polymer modified membrane is to be laid. The five layered waterproofing membrane shall be laid using Butane torch and sealing all joints and preparing the surface complete.

Drain outlets shall be properly finished to permit smooth flow of water in. Waterproofing treatment shall be carried into the drain pipe or outlets by at least 10 cm. The waterproofing treatment laid on the roof surface shall overlap the upper edge of the waterproofing treatment in the drain outsets by at least 10 cm.

The APP polymer modified prefabricated waterproofing membrane shall be cut to the required length. Waterproofing membrane shall normally be laid in length in the direction of the slope and laying shall be commenced at the lowest level and worked up to crest. APP waterproofing membrane shall be laid in 6–8 m lengths. The roof surface shall be cleaned and bitumen primer shall be applied in the correct quantity, over this specified waterproofing membrane shall be laid with butane torch after allowing 24 hours for primer to dry. Each strip shall overlap the preceding one by at least 10 cm at the longitudinal edges and 15 cm at the ends. All overlaps shall be firmly bonded with bitumen primer and levelled by heating the overlap with butane torch.

For terraces that are inaccessible, two coats of aluminium paint can be applied as APP is not resistant to UV rays. For accessible roofs the APP membrane need to be protected and this can be done with laying of cement concrete after sanding the APP surface. Alternatively accessible roof can be can be protected by brick tiles or broken white glazed tiles laid over 12 mm thick cement mortar of specified grade bedding and joints sealed with cement mortar of which shall be measured and paid for separately. White glazed tile roof finish helps in reducing the heat load of the upper floor on account of increased reflection of sun rays.

APP waterproofing membrane shall be laid as flashing wherever junction of vertical and horizontal surfaces occurs. Longitudinal laps shall be 10 cm. The upper edge of flashing membrane shall be well tucked into the flashing grooves in the parapets, chimney stack, etc., to a depth of not less than 6.5 cm; corresponding applications of primer coat shall also be made. The flashing treatment shall be firmly held in the grooves and it shall be sealed with the approved sealant after terminating the membrane.

Where parapet walls are of height 45 cm or less APP waterproofing membrane flashing shall be provided in the same manner as for splashing in the core of high parapet walls except that upper edge shall be carried out the full height of the wall and taken right across the top of the parapet and down on the external vertical faces to a minimum distance of 5 cm.

Where low dividing walls or inverted beams are met with, the same treatment shall be provided as for the main roof, the lateral bearing carried down both sides of the wall and overlapping the roof treatment.

Drain outlets, where formed in the low dividing walls, shall be given waterproofing treatment same as for the main roof.

Flexible liquid application of membranes manufactured by M/s WEBEC, KEMPER SYSTEM, Coniroof from M/s BASF can be used with advantage after checking economics.

7.13 ROOF DRAINAGE

Dampness and leakages from roofs are also common phenomena. As any leakage/seepage from the roof will aggravate the problem, certain salient points are discussed here.

1. Roof drainage should be planned in advance so as to have adequate number of drainage pipes of diameters as per design. Long runs of water on roof should be avoided.
2. Avoidance of long travel by water. Drainage outlets should be provided so as to ensure that the rain water does not have to travel distances longer than 15 m.
3. Sizes of rain water pipes should be adequate to ensure that there is no collection or heading of water at outlet. At mouth it is suggested that bigger diameter pipes are provided reducing into smaller vertical pipes.
4. Protection of inlets from floating matters such as plastic bags, leaves, etc. These prevent the ingress of water to rain water pipes. To ensure this gratings are provided at mouth of pipes and pipes which are vertical and flat with terrace are provided with vertical gratings/cowls so that floating plastic bag does not block opening totally. Of course regular cleaning of terraces cannot be over emphasised.
5. All waterproofing treatments must be taken along the parapet and secured so that the water flows only on terrace and has no chance of going underneath the waterproofing treatment.
6. Proper preparation of Khurras is essential so that the water drainage flows to well-treated outlet points.
7. During construction the tiles of mud phuska or other treatments should be properly grouted. Gola should be provided before plastering. Where tarfelts are provided adequate overlap should be ensured.
8. Quality of concrete for roof slabs requires special attention during execution. Concrete mix of M20 (1:1½:3) and above should be

used. Concrete should be well compacted. It is desirable to give slopes in top surface of concrete where feasible.

9. Roof surface to be provided with waterproofing treatment and finished with a material which is non-absorbent of water.

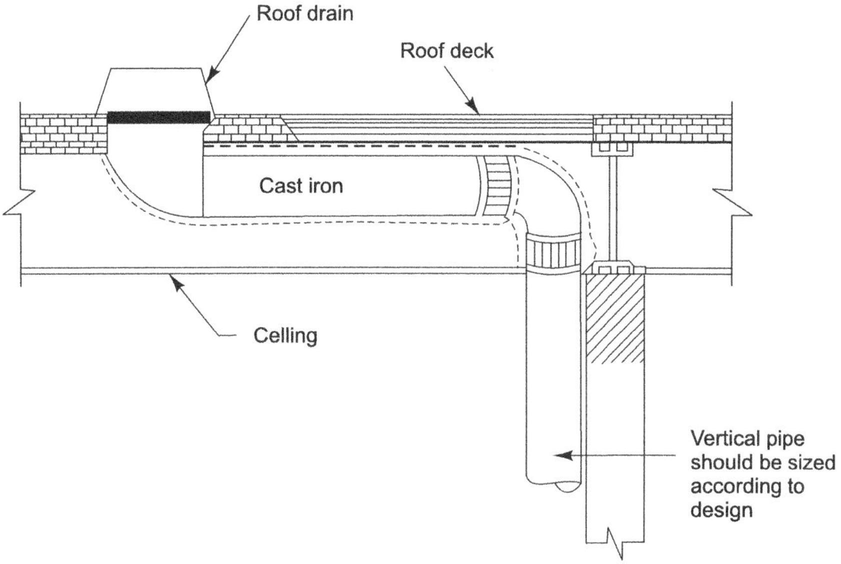

Fig. 7.6 *Typical roof drain*

10. In heavy rainfall areas or hilly areas sloped roofs should be provided. Roofs should project sufficiently from walls.
11. Adequate slopes (1:48) should be provided on roofs and unobstructed flow of water ensured. Accumulation of water on roof should not be allowed. Rain pipes should preferably be exposed.
12. Water tanks should be placed sufficiently above the terrace level to allow proper terracing below as also attention to maintenance jobs.
13. Preference should be given to locally available materials to facilitate maintenance later.

14. In case of roof slabs it is advisable to extend the roof slab by at least 300 mm from the outer face of the walls and the rain water pipes shall be provided vertically into the slab without bends near parapets.

15. No vegetation growth should be allowed.

16. Expansion joints should be designed and treated properly to ensure that there is no ingress of water.

7.13.1 Khurras

(a) The khurras shall be constructed before the brick masonry work in parapet wall is taken up and it shall be of size 45 cm ¥ 45 cm, and shall be made of cement concrete 1:1.5:3 mix (1 cement: 1.5 coarse sand: 3 graded stone aggregate 20 mm nominal size).

(b) Laying

A PVC sheet of size 1 m x 1 m x 400 micron (alternatively, aluminium foil of 32 SWG) shall be laid under the khurra and then cement concrete shall be laid over it to average thickness of 50 mm with its top surface lower that the level of adjoining roof surface by not less than 50 mm.

The concrete shall be laid to a size greater than the stipulated size of the khurra in such a way that the adjoining terracing shall overlap the concrete on its three edges by not less than 7.5 cm. The concrete will slope uniformly form the edges to the outlet, the slope being as much as possible, and in no case less than 20 mm cement concrete at the outlet. The concrete shall be continued at the same slope through the width of the wall into the outlet opening to ensure a water tight joint.

The khurras and the sides of the outlet shall then be rendered with 12 mm coat of cement plaster 1:3 mix (1 cement: 3 coarse sand). This shall be done when the concrete is still green and shall be finished. The sides of the khurras and sides of the outlet opening shall be well rounded. The size of the finished outlet opening shall be 10 cm wide and by 20 cm high.

Iron grating shall be provided at the outlet as a safeguard against choking. Iron gratings, shall be of overall size 20 x 25 cm with an outer frame of 15 x 3 mm M. S. flat to which 4 Nos M. S. bars of 10 mm diameter shall be welded in a vertical direction keeping equal clear spacing of 2.5 cm.

7.14 WATERPROOFING OF KITCHEN AND TOILET FLOORS

The floors of wet areas could be sunken or drainage pipes could be taken below the floor slabs. The waterproofing treatment will be designed accordingly. In both cases it is suggested that the structural slab is cast with good quality concrete and given a slope towards shaft. Imperviousness of the shaft should be checked by resorting to grouting.

7.14.1 Before the Waterproofing Treatment

Before the waterproofing treatment, the internal plaster of ceiling and walls leaving the portion for dado/skirting should be completed. Grooving/chasing for doing the concealed work of GI/CI pipes/electrical conduits should be completed. Cleaning the depressed/sunken portion of wet areas of all debris, extra mortar sticking to the vertical and horizontal surface, etc. should be done. Necessary holes for 'P' trap/ Nhani trap/Water escape pipe, etc. should be completed (Fig. 7.7).

7.14.2 Sunken Portions:

(a) Nomenclature of Item

Providing and laying waterproofing treatment in sunken portion of WC, bathroom, etc., by applying cement slurry mixed with Armourcrete of M/s The Structural Waterproofing Company Pvt. Ltd. or Tapecrete of M/s CICO Technology Limited consisting of applying: and the like consisting of:

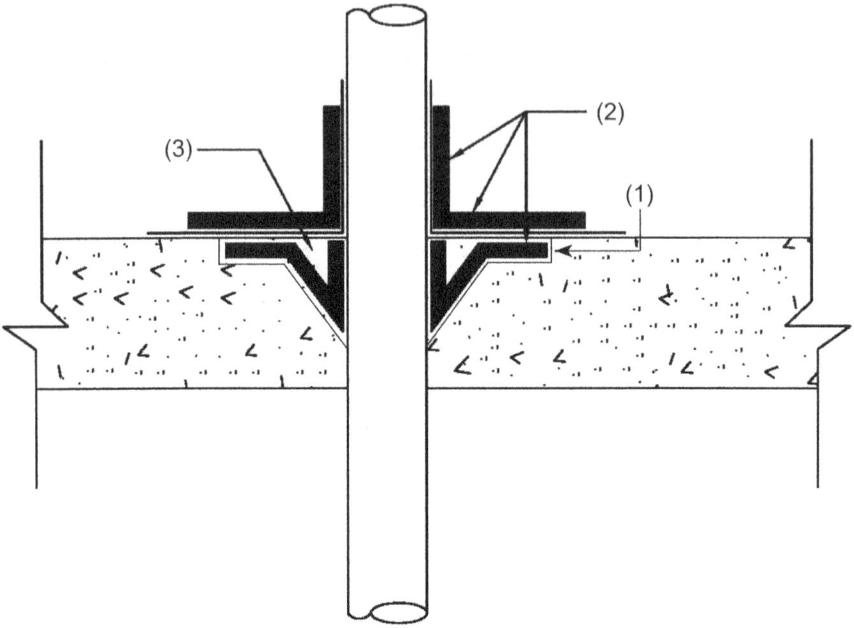

1. Cut U shape groove 25 mm wide 25 mm deep around pipe and clean
2. Apply crystalling waterproofing treatment into groove and pipe on damp surface
3. Fill groove with sealant (like tammspatch/duraltopgel from tamms)

Fig. 7.7 *Piping through floor/wall*

(i) First layer of slurry of cement @ 0.488 kg/sq m mixed with Armoucrete or Tapecrete @ 0.253 kg/sq m. This layer will be allowed to air cure for 4 hours.

(ii) Second layer of slurry of cement @ 0.242 kg/sq m mixed with Armoucrete or Tapecrete @ 0.126 kg/sq m. This layer will be allowed to air cure for 4 hours followed with water curing for 48 hours.

(iii) The rate includes preparation of surface, treatment and sealing of all joints, corners, junctions of pipes, and masonry with polymer mixed slurry. Rendering/plastering as protective layer, if required, will be paid separately.

(b) Methodology

(i) Surface preparation

Prior to application of TAPECRETE - P151 work, all surfaces must be prepared properly to avoid failure. The surface shall be cleaned to remove all dust, foreign matters, loose materials, or any deposits of contaminants which could affect the bond between the surface and the TAPECRETE-P 151 coating.

Use of sealing tapes which are Acrylic based can be made for expansion and construction joints under ceramic coverings. Acrylic based elastic packings are also available for sealing of pipe passages. To make the wall and floor joints waterproof below the ceramic tiles use of Acrylic based mouldings for inside and outside corners and joints are also available. One of the manufacturers is M/s Colossus India. Packings to seal water pipe openings are also available.

(ii) Laying

- New surfaces of the structural concrete shall be reasonably smooth to avoid sharp projections so as not to impede the application of TAPECRETE coating.
- All concrete surfaces shall be thoroughly pre-wetted for at least one hour prior to the application of TAPECRETE coating by sprinkling of water on flat surface and by vigorously spraying water on vertical/inclined surfaces.
- When placing TAPECRETE coating, all water shall be removed so that surface is only damp or surface dry. In no case there shall be standing water or a shiny wet surface.
- Depressions in floors are filled and levelled using PMC fillers. For the PMC filler the mixing ratio is 1 kg cement : 1.5 kg silica sand and 0.52 kg TAPECRETE P-151.

(c) Application

(i) First layer of slurry of cement @ 0.488 kg/sq m mixed with Armoucrete or Tapecrete @ 0.253 kg/sq m. The first layer shall be applied with painting brushes over the specified and dampened areas including corners, holes on the surfaces and joints of pipes in the concrete. And the application should continue at least up to 150 mm height of fixtures of pipes from the surface. This layer will be allowed to air cure for 4 hours (Fig. 7.8).

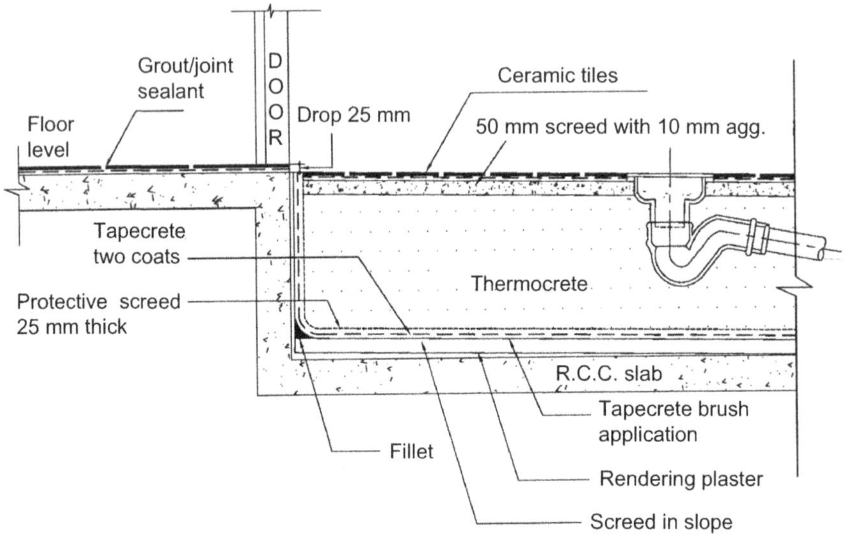

Fig. 7.8 *Waterproofing of sunken areas*

(ii) Second layer of slurry of cement @ 0.242 kg/sq m mixed with Armoucrete or Tapecrete P -151 Acrylic Polymer Modified Cementitious (PMC) @ 0.126 kg/sq m. Application in same way as specified in a) above. This layer will be allowed to air cure for 4 hours.

(iii) Application of one coat of TAPECRETE P - 151 - Acrylic Polymer Modified Cementitious brush topping over the PMC slurry applied surface. The PMC brush topping should be applied after the last PMC slurry coat has dried in the form of a 20 mm thick a 1 : 4 mortar (1 cement: 4 sand).

Providing protective overlay of 25 mm minimum thick screed concrete / plaster to slope of (1 in 100) admixed with CICO No. 1

Integral Cement Waterproofing compound after curing of PMC brush top coating. The surface then should be cured with water for 48 hours.

(d) Filling

The sunken portion should be filled with proper filling up to the desired level after placement of the required pipings with Thermocrete with styropore beads @ 18 kg per cum with sand and cement in ratio of 1:6 (Fig. 7.8).

(e) Walls

(i) Application of two coats of TAPECRETE P -151 - PMC slurry coating on the side wall over rendered surface and turning up to dado/ skirting level as per details given above.

(ii) Application of one coat of TAPECRETE P - 151 - PMC brush topping over the PMC slurry applied surface up to the dado/ skirting level.

(iii) Providing a 15 mm thick protective layer over the treated surface in the form of 1 : 4 :: Cement and Sand Plaster on floor and walls.

(f) Curing

(i) During the first 12 hours of curing, it must be protected from abrasion, rain and other adverse conditions.

(ii) No traffic shall be allowed on a standard TAPECRETE-P151 treated surface within 48 hours after installation.

(iii) After application of final coat of TAPECRETE-P151 composition, initial air drying shall be done for 2–6 hours. During this period no water is to be used for curing.

(iv) Moist curing shall be done for the next 24 hours by way of spraying water on the TAPECRETE system.

(v) Following moist curing, the TAPECRETE-P151 coating shall be allowed to air dry for 3 days before submersion in water.

(g) Flooring

(i) Lay minimum 50 mm thick concrete screed for laying the floor using 10 mm aggregate.

(ii) Use tile adhesives or cement mortar 1:4 for laying of floor tiles as per pattern and to slope.

(iii) Use waterproof tile grout for the entire tile joints.

(iv) Seal with butyl sealant all joints of the fittings to the floor as also the joints of the traps to the floor. Refer to Figs 2.12 and 2.13.

7.14.3 Precautions: Avoid the Following

(a) Use of tiles of the water absorbing type such a slate, etc.

(b) The practice of encasing the C.I. pipes in concrete, as the same offers no advantages whatsoever, and creates weak points.

(c) Filling up the sunken portions with cinder.

For rectification works steps to prevent ingress of water will depend upon the causes and recourse can be taken to:

7.15 WATERPROOFING OF UNDER GROUND TANKS

Two eventualities are to be taken care of:-

(a) Water table is high and causing hydrostatic pressure

(b) Water table is low

For Locations where the Water Table is High

(a) Treatment of integral waterproofing with rough Kota stone or APP membrane, as suggested for basement, needs to be adopted.

(b) Leave 15 mm diameter GI threaded nozzles in a grid pattern of 50 – 75 cm centre to centre on entire horizontal surface and at 50 cm c/c at construction joints.

(c) Similar nozzles shall be post fixed on critical points if required and on vertical walls of UGT.

(d) Injection of non-shrink polymer grouting compound admixed with cement through nozzles under pressure maximum 2.5 kg/sq cm by pump.

(e) Sealing of nozzles after injection operation with quick setting admixture with cement.

(f) Plaster the inside surface with a layer of cement mortar 1:3 with waterproofing compound finished neat and cured as per directions.

(g) Work to be executed by specialised agency, which should stand a warranty for 15 years for the work executed.

For Locations where the Water Table is Low

(a) Operation of fixing of nozzles and grouting to be same as given above.

(b) Catalytic in Depth Crystallisation system of Kryton or other manufactures as approved may be used for coating the inside surface so as to provide a flexible membrane barrier capable of bridging 1 mm gap width. It should be able to withstand water pressure of specified water head. The coating work shall be executed as per manufacturers'' specification after surface preparation as described above and will be executed through approved agency who will stand a warranty for 15 years. Brief steps are listed below:

(c) Surface Preparation and Krystal T1/T2 Application

- Concrete surfaces will be cleaned and made free of contaminants and laitance.
- Concrete must be sound.
- Concrete will be cleaned to have an open pore surface to allow penetration of Krystal. This may require mechanical preparation such as grinding, water blasting, sandblasting, or hacking.
- Surfaces to be treated will be pre-soaked with clean water to a saturated surface dry (SSD) condition. Do not leave any standing water.
- Mix Krystal T1 to a slurry consistency (5 parts powder to 2 part clean water). (Mix only enough material that can be placed in 15–20 minutes).

- Apply slurry to the concrete surface with a brush in a circular scrubbing motion so as to achieve slurry to the entire concrete slab. Krystol T1 with a spread @ rate of 0.8 kg/sq m.
- Mix Krystol T2 to be slurry consistency (5 parts of Krystol T2 to 2 parts of clean water) mix enough which can be applied in 15–20 minutes.
- Apply Krystol T2 slurry on the concrete surface where Krystol T1 treatment has been completed, with brush in a circular scrubbing motion so as to achieve maximum adhesion and penetration. The Krystol T2 spread shall be @ 0.8 kg/sq. m. This may be applied after 2–3 hours when Krystol T1 coating has set hard.
- Protect the area from rapidly drying out due to heat, damage by rain, excessive wind and freezing temperatures for 48 hours.

Joint Treatment

For construction joint treatment a U-shaped groove will be made of 10 mm wide and 15–20 mm deep. This will be filled up with Krystol T1 putty consistency 5:1 (5 parts of T1 to 1 part of water), and Krystol Baricote in 3.5:1 (3.5 part of Baricote to 1 part of water) consistency as per manufacturer's specification. First half of the joint is filled by Krystol T1 and balance portion by Krystol Baricote.

(a) The coating shall be protected with a layer of cement mortar with waterproofing compound finished neat and cured as per directions.

(b) Testing After Construction

(i) It is detrimental to keep the water retaining structure dry for a period longer than four weeks, as it may lead to formation of cracks. So it is imperative that before the last casting is completed, water arrangement for testing the tank is ready at site. Immediately after the removal of form work, the tank should be tested. All preliminaries should be completed in advance.

(ii) Water should be supplied to the water tank slowly at the rate of 300–450 mm depth of tank per day and the result closely observed. At the end of the operation, that is when full supply level is reached, all valves shall be closed tightly. The water level in the water tank should be properly marked on the wall. Leakage through the valves should have been checked and there should not be any drop due to the same. After 24, 48 and 72 h, the levels should be checked and the drops in level will be a measure of water tightness.

(iii) The permissible standard usually adopted is 6 mm drop in 24 h in case of covered water tank and 12 mm in case of open reservoirs. If there is no drop but dampness is observed in the outer surface, such dampness may vanish in course of time as the free lime ejected out of cement will be plugging the minor pores causing such dampness. If the intensity of leakage is slightly more, then lime may be added to the testing water.

In case of leakages, the points should be marked and separately treated after dewatering.

(d) As an additional precaution, exterior of tank shall also be given a coat of plaster with water proofing compound mixed in cement mortar 1 : 4 (1 cement : 4 coarse sand).

(e) For remedial measures, after construction refer 11.3.

7.16 WATERPROOFING OF OVERHEAD TANKS

(a) Injection grouting to be done in a manner similar to UGT.

(b) Catalytic in Depth Crystallisation system of Kryton or other manufactures as approved may be used for coating the inside surface so as to provide a flexible membrane barrier capable of bridging 1 mm gap width.

(c) Inside surface will then be plastered with Cement mortar 1:3 mixed with waterproofing compound.

Other precautions are to be on the lines mentioned above for Under Ground Tanks.

7.17 NEW METHODS

It is important to choose right products and system to achieve a durable and impermeable concrete structure which can last more than its anticipated life. A number of chemicals, sealants and membranes/ treatments are now available for application of the wet areas and stopping water ingress. A list of some of these materials with their manufacturer's names and brief details and uses is appended in Annexure 'B'. While using these products, the manufacturer's recommendations must be kept in view. For specific applications, selection of method/material has to be carefully decided.

CHAPTER 8

TESTING AND INSPECTION

Testing of pipelines, fittings and fixtures should be carried out in stages. As soon as materials are received, certain tests to check the quality, such as abrasion test for tiles, visual tests for seeing that first quality G.I. pipes and sanitary fittings are to be conducted.

After pipes are laid and joints are made, such assembly requires to be tested whether it can withstand water pressure that it will carry, and also to see effectiveness of the joints. Water system should be able to withstand a hydraulic pressure of 6 kg/sq cm without showing signs of leakage. The hydraulic performance of fittings and cisterns should be seen. The pipes should be slowly and carefully charged with water allowing all air to escape and avoiding all shock or water hammer. The draw off takes and stop cocks will then be closed and specified hydraulic pressure applied gradually. Test pressure should be maintained without loss for at least half an hour. The testing can go section wise as work proceeds.

No walls, chases, floors should be covered till the lines laid are tested as above. After all lines and fixtures are connected, the system should be tested in accordance with the provisions of IS 1742:1983. Only when the system is found to be satisfactory in the test it should be allowed to put to use.

8.1 SMOKE TEST

All traps are filled with water before starting the smoke test. A smoke testing machine consists of a length of flexible rubber tubing and bellows. The smoke is made by firing oily waste (brown paper or cotton waste soaked in creosote) which gives a clear pungent smoke, easy to detect by sight and smell, if there is any leakage at any point. Smoke is pumped into the drains and pipes at the lowest end through a gully outside the house

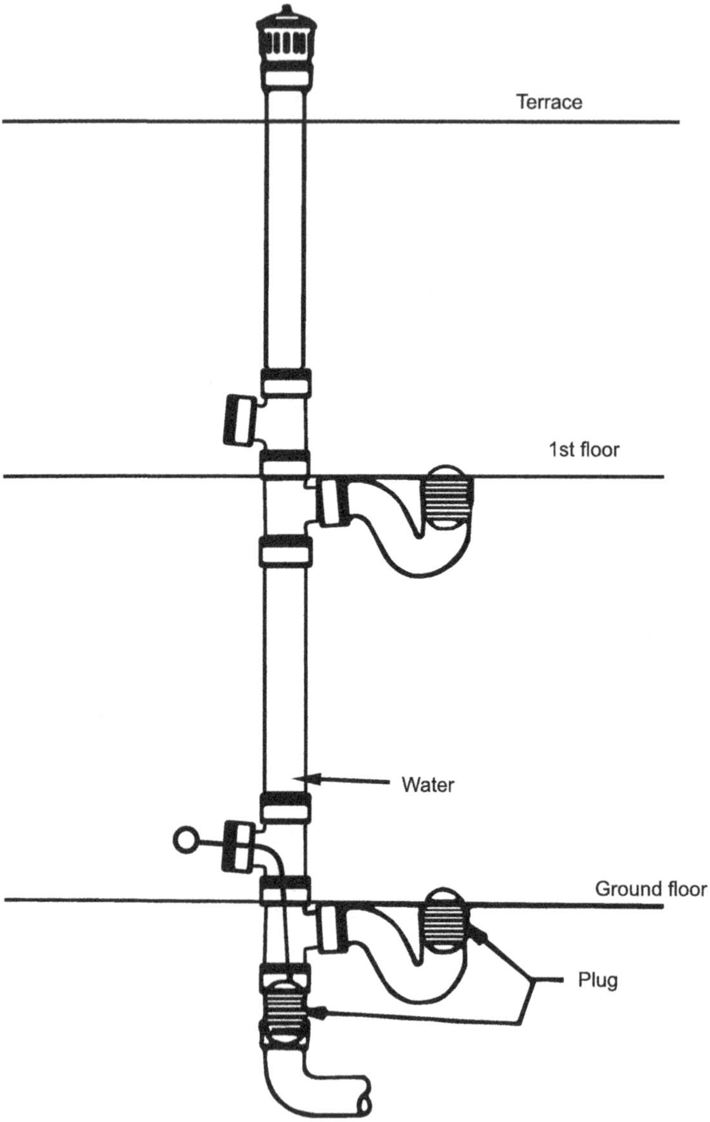

Fig. 8.1 *Soil pipe testing*

or an inlet ventilator, or through a clay plug in an inspection chamber. In making a smoke test, the tops of soil and ventilating pipes are left open until smoke is seen to issue when the openings are plugged securely with wet cloth or wet clay tied in cloth, and smoke pumped in for some considerable time.

8.2 WATER-TEST FOR STONEWARE AND CONCRETE PIPES

After the joints have properly dried (for at least seven days) and before filling the trenches, the pipes should be tested for water-tightness by filling the pipes with water to the level of 2.5 m above the top of the highest pipe in the length to be tested, by closing the ends of the sections and maintaining this water level for one hour. Earthenware pipes should not be subjected to a head of more than 3 m of water. A plug is inserted at the lower end of each length and a right angled bend at the top and funnel fixed through a rubber tube or testing rubber plug is used. (A drain plug is cylindrical bag of rubber and canvas to which a tube and a tap valve is fixed at one end. Air is pumped into the plug which is inflated and blocks the passage of water. If these plugs are not available use of wad of clay supported by a disc of wood can be made). After air bubbles have escaped following the first filing and absorption has ceased, water is again added to completely fill the pipe.

A light amount of sweating that is uniform may be overlooked, and a small amount of subsidence should not be taken as implying bad workmanship or defects. Absorption is at a diminishing rate of subsidence till no further subsidence takes place.

The water put in the pipes for testing should not be drained out until the trenches have been filled in about 90 cm to detect if any joints have given way during the filling. Alternatively, the test should be repeated after back-filling the trench.

Branch drains having a trap at their upper ends, unless they are provided with a cleaning of ventilating branch, become air-bound, so that the results of a water-test cannot be accurately gauged. To obviate this, the confined air should be drawn off by means of a bent tube.

8.3 WATER-TEST FOR CAST IRON DRAIN PIPES

In the test for water tightness of the joint, the ends are closed by flanges. A small diameter pipe is inserted in the upper end and a valve for escape of the air in the other end. The pipe is filled with water until water stands in the small pipe to the required height. The water pressure should be maintained for not less 10 minutes. This method is used for heads up to about 6 m. Hydraulic pressure pump is also used, until gauge shows required pressure. If there are no leaking joints, the water level in the pipe or the pressure on the gauge will be maintained. Pressure pump is used for test of higher pressures than 6 m of head.

8.4 TESTING OF CPVC AND PPR PIPES

The most common test method is to use water under moderate pressure. Air testing is not recommended as pipe and fitting manufacturers do not recommend air testing, and cannot be held liable for any injuries occurring during the air testing of their product

8.5 RECOMMENDATIONS FOR WATER TESTING

Water testing is a safer, more reliable and more accurate method for testing plastic piping systems. Because PPR and CPVC pipe and fittings are designed to convey liquids, most companies recommend testing with water. The purpose of the test is to locate any leaks at the joints and correct these prior to putting the system into operation. Because it is important to visually inspect the joints, a water test must be conducted prior to closing in the piping or back filling underground piping.

If there is a leak in the system, it will always be easier to locate when testing with water; air leaks can be hard to find.

To properly water test, plugs should be inserted through test tees to isolate each section being tested. All other openings should be plugged or capped with test plugs or test caps. Then fill the system being tested with water to the highest point. In the pipeline the pressure reaches 7–8 kg. Because the daily water consumption, pressure is 3–4 kg. Then look at the pace of decline in pressure gauge, and then check for any leaks in pipe joints.

If a leak is found, the joint must be cut out and discarded. A new section can be installed using couplings. Once the system has been successfully tested, it should be drained and the next section should be prepared for testing.

CHAPTER 9

MAINTENANCE AND HOUSEKEEPING

Maintenance aims at effective and economic means of keeping the building and services in full utilisation. Maintenance activity continues for the entire life of the structure and has lot of bearing on the life of structure. It makes significant effect on the satisfaction levels of occupiers and users.

Maintenance is an art and is more important than construction. Maintenance could be preventive or corrective. One has to analyse the cause of the problem, frequency of occurrence and its effect on the users/ occupants. Failures of fittings/fixtures or systems should be studied to take corrective measures. Works which are carried out in anticipation of failure are termed as preventive maintenance and those which are taken to correct the faulty appliance or service as corrective maintenance.

Planning for good maintenance has to be done even before execution. Design of various areas has to be such that all appliances and areas needing frequent maintenance are easily approachable. Adequate working space should be available to maintenance personnel.

Since we are discussing wet areas in particular, emphasis is on having working platforms with easy access in shafts. Similarly terrace should be accessible to the attending staff. Specific provisions are required to be made for maintaining services in multi-storeyed buildings.

Maintenance should be done in regular basis and one should not let things deteriorate to an extent that cost of repair may be more than the original cost. Hence emphasis should be on preventive maintenance.

Proper maintenance would ensure that all fittings and fixtures are kept in acceptable working condition so that their utility is not sacrificed.

Checking of roofs, drains, traps on regular basis to ensure they are clean and functional is part of preventive maintenance. Users and cleaning personnel have to be educated not to throw solid waste into WC's, traps and drainage system.

It is good practice to lay down a periodical drill for doing some of the maintenance works required to be done in regular basis such as white washing, painting, drain de-silting, roads re-carpeting, cleaning of traps, manholes, cleaning of tanks, etc. A small inventory of items which are often required should be kept in store for immediate use.

For effective maintenance it is necessary that the buildings are inspected on regular basis and corrective measures taken as per the outcome of inspection. Special attention is required to be paid to inspection of storm water drainage works, roofs before monsoon.

The maintenance of water supply and sanitary system in buildings requires thorough knowledge about the design, specifications and the materials used within the building and for external connections. If the defects are not attended to timely these may result in ultimate structural failures too. Availability of as built drawings for services is of great help in taking up repair and restoration works. Maintenance manuals should be got prepared for each building at the time of construction is getting completed.

9.1 MAINTENANCE SURVEY

Even the best building ever constructed will get ruined in due course of time if not properly maintained. A delay for sometime is tolerable in case of buildings, but in case of services it can lead to bigger hazards, including danger to building life. Therefore, regular and periodic surveys are necessary for maintaining building service ability. Proper inspection is time consuming job. Proper way to start inspection is to obtain drawing according to which services have been laid. All service lines should be shown by different colours for systematic checking and inspection. Site for inspection must be prepared for testing of various services. Prepare a check list for specific inspection and necessary test schedule. The checklist should be comprehensive, systematic and simple to follow. All occupants must be informed of the inspection and report making for the purpose of maintenance of services. The site must be available and accessible for inspection and testing. A comprehensive report should be prepared after inspection, testing and interviews with the users of services. A sample check list is given as Annexure C.

9.2 MAINTENANCE OF PIPES

For repair jobs, a plumber should carry a repair kit. Repair kit should contain rubber padding, clamps, spanner, pipe wrench, sockets, nipples, sealing paste, stoppers, union, etc. The pipe line requiring replacement of any section must have union for easy and quick replacement. While designing water supply lines, provision of unions facilitate repair and replacement.

9.3 REPAIR AND MAINTENANCE OF OVERHEAD AND UNDERGROUND WATER TANK IN BUILDINGS

Overhead tanks are provided in buildings to give uninterrupted water supply to the occupants. The size of tank depends upon the number of occupants, per capita water requirement and number of water closets it has to serve. The tanks may be constructed from different materials and in different shapes. The most common shapes are rectangular and circular. The outlets from the tanks should always be provided with anti-siphonage pipe to avoid air locking.

The OHT and UGT must be periodically checked for any leakages, proper functioning of float valves. Regular cleaning of tanks must also be ensured at regular intervals of six months and cleaning date must be marked.

9.4 MAINTENANCE OF APPLIANCES: TOILETS

A clogged trap in a water closet requires careful handling as water closets are made of vitreous china which might crack if exposed to extremely hot water or excessive force.

A plunger will normally handle simple toilet clogs. Another method of cleaning a water closet trap or toilet is the use of an auger with an adjustable, crank-type handle. Known to plumbers as a "snake," the spring-steel coil is easily worked past the trap and down the pipe.

If the rubber-cupped plunger or the auger does not clear the toilet, then professional help should be sought.

Tubs

When trying to clear a plugged bathtub drain, place a heavy cloth in the bottom of the bathtub to avoid scratches on the bath's enamelled surface. Hold your hand or rag over the waste and overflow plate, cup the plunger over the drain and plunge it vigorously several times. If it doesn't open easily, the drain may require cabling to open it.

Heavy steel spring coils should not be used to clean traps under lavatories, sinks, or bathtubs. A more flexible type of wire or spring should be used—one which is easy to work through the bend of the trap (Fig. 10.2).

Water Geyser

For gas or electric water heater, set the temperature dial at or below the suggested factory energy savings settings listed on the water heater. Above that mark means excessive wear on the water heater and the potential for scalding.

9.5 HOUSEKEEPING

No installation would serve its desired purpose if it is allowed to deteriorate due to neglect or lack of proper upkeep. One would like to have fittings and fixtures sparkling clean, functional and pleasant to look at. Bad odours would occur if the installations are not well looked after on daily basis as organic matter would putrefy. It is therefore essential that lavatories and all wet areas should be cleaned thoroughly on regular basis.

It should be remembered that the pipe installation is a close system and therefore if trash and debris gets into the pipe fixtures, it is very difficult to get it out. Vegetables pieces, piece of bones, bidi/cigarette butts, waste cloth and polythene bags; and in hospital, medicine bottles, lint, gauge, cotton, injection syringes, etc., find their way to pipe lines when proper facility for solid waste disposal are not provided. In public premises bins, etc., have to be made vandal proof.

A regular drill for chores related to housekeeping to be attended on daily, weekly and monthly basis can be laid out and followed to ensure good up-keep, just like for maintenance works. Good and efficient housekeeping and its awareness will not only give hygienic environment, but will result in longer life to fittings and fixtures and comfort to users, and may reduce costly maintenance.

9.6 ACCESSIBILITY FOR MAINTENANCE AND HOUSEKRRPING

While planning services, maintenance of these services should be kept in mind. Vent doors and pipes should be available for inspection and easily approachable. Adequate clean outs are to be provided.

Working platforms need to be provided, especially in shafts, to facilitate maintenance of various services. Such platforms should have easy access from inside of the buildings. Normally access to such platforms is given from service areas which are connected to the shafts e.g. Toilets. Location of such platforms is important. These have to be provided at a level where from a person can attend to various joints, bends, fittings and fixtures and has safe working place. Hence these will always be at a lower level than the location of joints, bends, fittings and fixtures. These platforms should be sturdy and safe requiring minimum maintenance themselves. These can be fabricated with materials such as Steel sections and IRC fabric to provide light weight and economical structure.

9.7 PRE-MONSOON CARE:

Instances of leakages/seepages show sharp increase during monsoon season. Hence, special attention is to be given to pre-monsoon care of the following:-

(a) Terraces

Keep the terrace absolutely clean of any debris, waste material, etc. Any visible defects should be repaired.

(b) Mouth of Rain Water Pipes

Ensure that this is not choked and is clean of earth, vegetation, etc. It is commonly observed that horizontal bend of the Rain Water Pipe is not pulled beyond the inside face of the parapet. As a result, a portion of the water enters the building and causes deterioration.

(c) Sub soil water rising by capillary should be attended before monsoon.

(d) Cracks, separation and gaps Seal these with acrylic polymer mortar.

(e) Hair cracks

Dampness is experienced from the hair cracks on top of the parapet, in the parapet plaster and external plaster. These cracks should be filled with an acrylic-based flexible putty.

CHAPTER **10**

COMMON PROBLEMS

The faults in the services occur due to defective system and misuse. The common problems in water supply and sanitary system are as follows:

- Leakage in internal pipes
- Blocked drains and leaking joints
- Floor drains
- Overflowing cisterns
- Blocked waste pipes
- Fittings
- Inadequate pressure
- Maintenance of appliances
- Noise in plumbing system
- Odour in piping system
- Seepage through floor joints of wet areas
- Heading up of water on terraces
- Use of improper materials fittings/ textures.
- Damaged or cracked appliances

10.1 LEAKAGE IN INTERNAL PIPES

If wet patches are noticed on ceilings or walls at all the times, then these are indicative of leakage from pipes. Location of maximum dampness points to the most likely source/location of leakage. The dampness near C. I. fitting is caused due to leakage through C. I. fitting. Seepages are commonly seen at junction of floor and the 'P' trap or the floor trap. This problem can be avoided by proper sealing of gap between floor and the trap at the time of original construction as detailed in execution stage.

If a leak is noticed in the plumbing system the water supply should be shut off and plumbing contractor called at once. Water supply systems are under pressure. Temporary repairs are only temporary and wrapping the pipe usually fails. Leaks must get immediate attention.

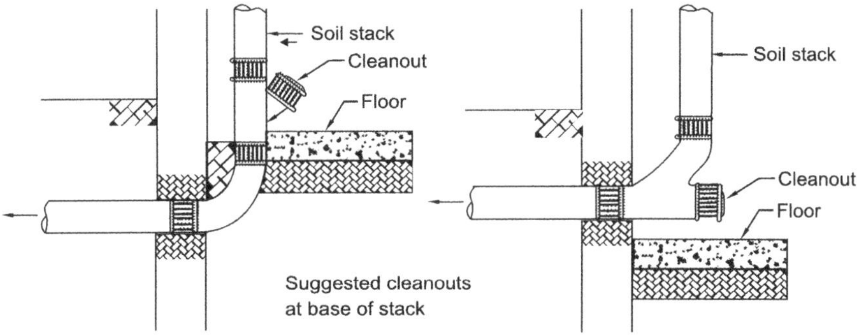

Fig. 10.1 *Suggested cleanouts using cast iron soil pipe*

10.2 EARLY DETECTION OF LEAKAGE AND SEEPAGE

Normally the leakages and seepages are noticed after the damp has saturated the surface and patches/marks are visible to naked eye. To take action as soon as the problem starts, early detection is of utmost importance. Internal leakage can be checked by closing all the taps and observing the movement of the meter needle. If it is moving then it will clearly indicate internal leakage in the line, which can then be located and set right by replacing defective section. If it is in walls or roofs, it will dampen the wall or the roof. First check the pipeline underground for leakage and then check rest of the pipe under the floors and locate correctly. Locating this defect and setting it right will also save the structure from damage, as the leakage of pipes under the floor will result in seepage of water directly to the foundations.

Now-a-days, thermal cameras are available which can detect the early signs of leakages, etc., and can actually photograph the same showing location and its intensity. These can be made use of before the issue of completion certificates for the buildings, for taking up of repair works and also seeing effectiveness of repairs so carried out.

10.3 BLOCKED DRAINS AND LEAKING JOINTS

Clogged drains account for the maximum calls to plumbers/ serviceman than for any other service. The most-used drain is the one in the kitchen sink and that is the drain most often clogged (Fig. 6.3).

Preventing this situation can be done by carefully watching what is emptied into the sink drain and by the regular use of a safe biodegradable waste digester.

Sink stoppages are usually caused by liquid fats, emulsified by warm dishwater and carried through the pipes. Pour excess grease into a tin can and throw it out with the garbage, not down the sink drain. When using a food disposer, always let sufficient cold water run to carry the particles down and into the main line to prevent build-up in the smaller waste lines.

In the event of a stoppage, you should have a "plumber's friend," or plunger—a large rubber suction cup with a wooden handle. Cup it tightly over the drain and plunge it vigorously several times.

Drains may get clogged by large size objects dropped in to them or by accumulation of grease, dirt or other matter. If the obstruction is in a trap, the trap cover can be removed and cleared. If the obstruction is elsewhere in the pipe, cleaning can be done by using extendable rods attached with flushing rubber cups in front. This rod should be moved to and fro in the clogged drains and flushed with water. If the clogging is more hard and difficult, use augers with steel cable.

Cleanout augers with long, flexible, steel cables, commonly called snakes, may be run into drain partly to break the obstruction and to hook on the pull out the obstructing objects. Augers are available in various lengths and diameters and should be selected according to the size of the drain.

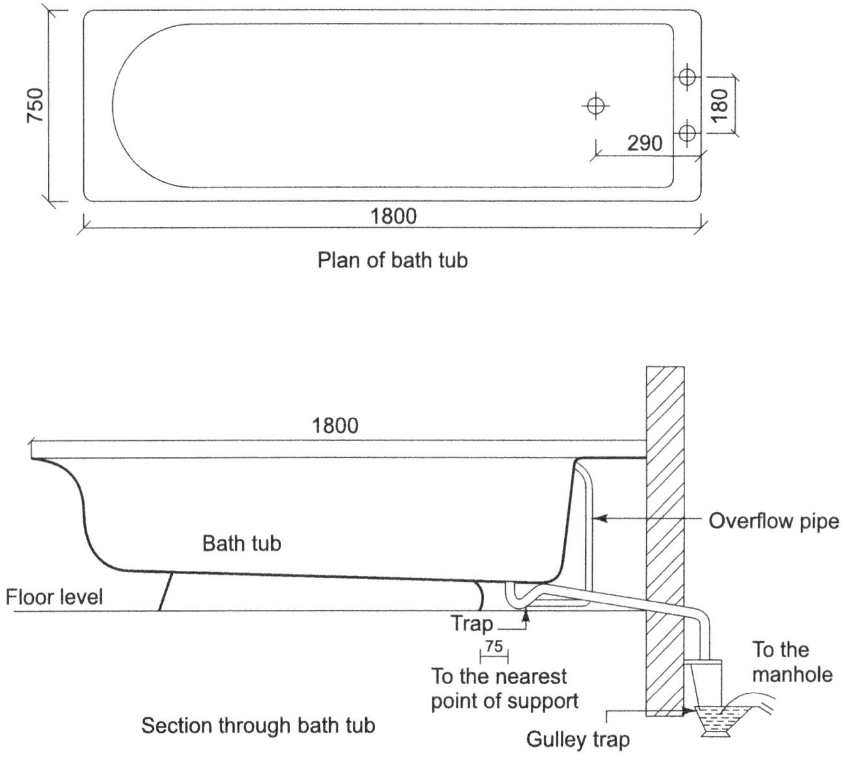

Installation of bath tub showing the position of inlet, outlet and trap

Fig. 10.2 *Installation of bath tub showing the position of inlet, outlet and trap*

Small obstructions can sometimes be forced down or drawn up by use of an ordinary rubber cup force. Flushing of grease and soap with hot water may be carried out for 10 minutes. Sand and dirt sometimes clog the floor trap and drains. Remove the strainer and take out much of the sediment around the floor trap manually and flush the drain with clear water. If more pressure is needed, use a garden hose to clean the clogging. Push and fix warp clothes around the hose into the drain to prevent flush back flow of water.

The floor drain should be checked regularly, especially one that is not often used, since water in the trap may evaporate. This would allow sewer gases to enter the room. Pour a bucket of water into the drain periodically in order to ensure proper water seal (see Fig. 10.3).

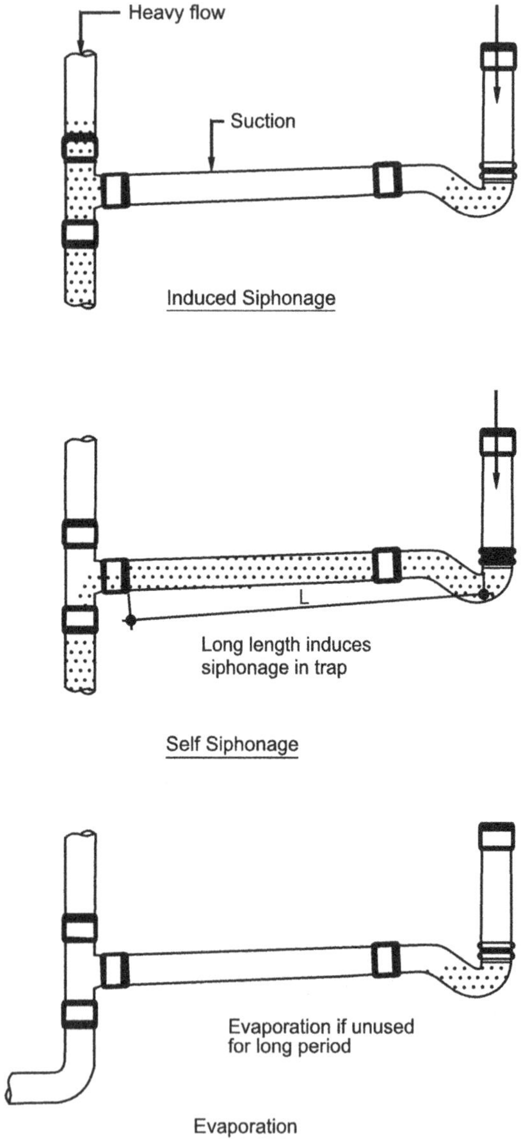

Fig. 10.3 *Reasons for loss of trap seal*

10.4 WC CISTERNS

Generally a cistern part gets deteriorated and requires frequent repair and setting. Parts that usually require repair are flush valve, the intake valve and float. The problem is either of water flowing continuously or not flowing at all. Overflow from the cistern or the overhead tank occurs due to a damaged ball cock or due to blockage of the ball valve itself. The ball may develop a crack or leak, or get disconnected from the copper rod. This requires replacement of the ball or the whole ball assembly.

Fault can be due to the presence of some grit in the valve, damaged washer or erosion of the seat, the grit can be removed, the washer can be replaced and if necessary the defective seating can also be replaced.

Sometimes WC does the flush which indicates that bell type or siphon type arrangement is out of alignment. The bell which induces the siphon, jumps out of seat which requires resetting and placing back in position. When siphonic type flushing cistern is either broken or worn out severely, the cistern should be replaced.

10.5 BLOCKED WASTE PIPES

Blocked waste pipe results in the delayed/ slow emptying of the fitting to which the water pipe is connected. Normally cleaning eyes/plugs are provided in the pipe for cleaning and the same should be opened to remove foreign matter. (Fig.10.1) To decrease the chances of the occurrence of this defect, a strainer should be fixed at the entry point of waste pipe. Flow should be checked in the waste pipe at regular intervals.

10.6 FITTINGS

Fittings (faucets and valves) are used more often than any other part of the plumbing system. The best modern fittings are all chrome plated brass and will last a lifetime under everyday use. They clean easily with soap and warm water.

When a tap does not completely close to stop water, there are three possible causes:

- A defective handle not capable of pressing washer properly against the seat.
- The deteriorated or broken washer not capable to block the opening properly.
- Seat is worn out or broken.

A steady drip which comes out drop-by-drop can waste as much as of 9000 litres of water every three months. A good quality tap has a life of about 10 years and in most of cases it is the washer which deteriorates and its replacement cost very little. Defective tap results in noise when water is running.

Washerless Faucets: When repairing washerless faucet or requesting service on one, it is vital that you know the brand name, or have a sample of the part you require, as there are varieties of faucet cartridges and parts kits on the market today.

Compression Faucets: In a compression type faucet, replacing the washer usually will correct a dripping faucet. When washer is to be replaced, it is enough to take out the upper part of the tap along with the washer spindle assembly. Fix new washer to the spindle assembly with screw.

However, when removing the stem, always check the seat inside the faucet body—the brass ring that the washer grinds against. The faucet seat can be worn or grooved, making the washer replacement ineffective within days.

Some of the newer, soft neoprene washers are for both hot and cold water and have a long life. The washer should fit snugly without having to be forced into position.

The metal chromium is easily dissolved in hydrochloric acid and sulphuric acid. Diluted hydrochloric acid has been considered a good tile cleaner, but only where there are nickel plated plumbing fittings. Where chrome plating is present no acids should be used and clean bathroom tile with warm oxalic acid never with aqueous solution of HCL or sulphuric acids.

10.7 INADEQUATE PRESSURE

Low pressure may be the result of undersize pipe, blockage in the ferrule, meter, main stopcock, or wrong fitting of stopcock and meter. These have to be checked and set right when found blocked.

10.8 NOISES IN THE PLUMBING SYSTEM

In designing the plumbing system for a new house, a plumbing contractor should endeavour to make it as noiseless as possible.

Much of the noise is due to water travelling at a high velocity, it follows that reduction in velocity of the water will correspondingly reduce the noise in the system. Larger pipe will not only provide a more adequate supply of water but also will reduce noise.

There are three general types of noises found in plumbing systems. These are water hammer, whistling and chattering.

Water hammer is the thump in the piping heard when faucets or valves are turned off abruptly.

Water hammer should not be permitted to go on indefinitely. The noise is only an audible symptom of what is going on in the piping. The piping is being subjected to the wear and tear of a multitude of shock waves. The result will be leaks in piping, tanks or fixtures unless the condition is corrected. Supply system should be provided with proper air venting.

Chattering in the piping may be caused by loose pipes, by pipes rubbing against a metal projection, by worn faucet washers or looseness of other inside parts.

Whistling is caused by the speed of water flowing through piping which is usually too small. A pressure reducing valve will help as will a general straightening out of the plumbing system. Whistling is most common at bends and tees in the pipe.

10.9 ODOURS IN THE PLUMBING SYSTEM

The well-designed and correctly installed plumbing system is odourless. Odours are most likely to arise from leaks in the waste or vent piping or from traps which have lost their water seal.

Unusual odours should never be ignored. Such odours are often an indication that sewer gas is present.

Every plumbing fixture is the terminus of the city water supply system and the beginning of the city sewerage system. The faucets control the water supply. The traps and vents control the sewer air. Sewer air is prevented by water barrier. In a trap, which is plainly visible under such plumbing fixtures as sinks and lavatories.

10.11 WHERE AND HOW TO SHUT OFF WATER

Knowing where and how to shut off water for the entire house or any part of it can be mighty important in an emergency. That is why it is extremely important for all members of the family to know where the valves are and in which direction they should be turned to shut off the water.

The most important valve in the house is the main shut-off valve for the entire plumbing system. This valve, generally located on the house side of the water meter, usually has a handle like a wheel it is advisable to place a few drops of oil around the valve handle once or twice a year. This will prevent the sticking action of corrosion.

In addition to the main strut-off valve at the meter, the well plumbed house has individual shut-off valves on the branch lines leading to individual fixtures, groups of fixtures or equipment such as water heaters, water softeners, automatic washers, etc.

10.12 DAMAGED OR CRACKED CHINAWARE

Damaged chinaware fittings give a very shabby look in addition to the leakage problem. Chinaware fittings can only be replaced if these are cracked. Washbasins are mostly fixed on iron clamps embedded in wall up to very small depth. A slight displacement can break them. The support angles should invariably be embedded adequately (200–250 mm) in the wall so as to make them more stable. Whenever supporting fixtures are loose, these should be immediately repaired and reset.

10.13 DEFECTIVE DOWN TAKE DRAIN PIPES

Down take drain pipes should be non-leaking. Metal pipes are favoured as these are non-perforated and non-leaking. Remedial measures will consist of painting of the inside of gutter and downpipes with bitumen paint. Otherwise these pipes may have to be replaced.

In case of plastic pipes, bowing during hot season is major problem. Generally it is reversible. However, if joints start leaking, these need to be repaired by using appropriate leak proof sealing adhesive. Bowing of plastic pipes can be straightened at average temperatures and joints checked for leak proofing.

10.14 SEEPAGE THROUGH FLOOR JOINTS OF WET AREAS

Joints of walls and floors and within floors have to be done in proper sequence and then sealed adequately to prevent water ingress through them. Joint details have been explained in detail in section and may be followed. Joints need to be sealed with good quality joint sealants and should be inspected periodically to check their effectiveness. Special attention is required while sealing joints of floors and fixtures.

10.15 HEADING UP OF WATER ON TERRACES

Designing of terrace drainage is of utmost importance. Keeping in view the rain intensity, area of roof, number and size of drainage pipes is designed so that there is no accumulation of water on terraces. Correct execution of Khurra also helps in faster disposal of rain water. As mentioned in monsoon care drain inlets have to be kept free from blockages.

CHAPTER 11

RECTIFICATIONS

If the required preventive measures have not been taken in the initial stages or have proved to be inadequate, then one is faced with the actual problems that are outcome of lack of such measures. This could be in form of dampness on walls, ceilings or floors, ugly patches on exterior and/or interior of walls, peeling of paint/finishes, swelling of timber members, bad odour and unhygienic conditions. Rectification measure are then required to be taken for which sources of leakage and seepages are first determined and treatments provided to prevent ingress of unwanted water. Such rectification costs are quite high and many a times their implementation costs and also cause inconvenience to users.

11.1 LOCATION

Once leakage/seepage has set in after the building is constructed, only option left is to resort to rectification measures and do repairs in a way that the problem does not recur. Before, however, any rectification measure is adopted, it is necessary to locate and determine the cause of leakage/seepage. This more often than not is not a simple affair. The cause can be at some place but damage is seen at a faraway place. To locate the fault or source following methods are adopted:

(a) Visual inspection by an experienced Engineer/technician to see the locations and likely sources of ingress of water.

(b) Study of original as built drawings for services and specifications adopted.

(c) Mark cracks for their location and status and see if these are increasing or remain same.

(d) Life of pipes, fittings and fixtures and remaining useful life vis-á-vis life of building.

(e) Signs of fungus/algae which indicate water ingress.

(f) Early signs and symptoms of seepage/leakage can be seen through modern detection methods involving photography of hidden areas.

(g) Detection methods giving photographs of occurrence of dampness are also available.

11.2 RECTIFICATION WORKS

Depending upon the location and identified cause for ingress of water, rectification strategy is to be worked out. This could involve:

- Sealing of cracks
- Stopping water ingress by using foaming agents.
- Providing or redoing waterproofing system. Various products available with their suggested use areas are listed in Annexure B and can be made use of in consultation with manufacturers.
- Injection grouting to stop the entry of water
- Seal the leaking joints using sealants
- Apply waterproofing chemicals/membranes where the old one has given way or is non-existent.
- Provide Damp proof course to stop rising of water through capillary action. New DPC layer can be inserted in existing buildings.
- Improve drainage of surrounding areas if found to be inadequate.

11.3 REMEDIAL TREATMENT

(i) The nature of the remedial treatment will depend on the extent and magnitude of the leakage noticed as given below.

(ii) In case of leakage through the opening of the horizontal construction joint, after emptying the reservoirs, the leaky joint should be properly cut-off preferably by using a power-driven carborundum saw or any other tools so as to get a square or slightly undercut groove. The joint is then filled up with cement mortar or synthetic resin after coated with mastic or bitumen.

(iii) Coating with sodium silicate- The solution made of one part of sodium silicate commercial (sp. gravity 1.8 kg/litre, paste variety of purity containing sodium oxide and silica in the ratio of 1:2.6–1:2.8) and three parts of water by volume is applied copiously with a soft broom over the entire surface. The treatment consists of three coats applied at an interval of 24 h between each coat.

(iv) Chemical Injection- In the case of severe leakage, this method as indicated below, may be carried out. The main advantage of this method is that this treatment may be applied at any stage, that is, even if leakage or the seepage problem develops long after the construction.

- Holes about 50 mm diameter and 25–40 mm deep shall be chiselled in grid pattern at a spacing not exceeding 1.5 m centre-to-centre all over the base, walls and top slab. In additional all the construction joints shall be opened by making a groove as to reach the reinforcement.
- MS grouting nozzles of 25 mm diameter shall then be fixed in these holes and grooves. After the nozzles are fully set, neat cement slurry admixed with water soluble monomer based chemical shall be injected through the network of nozzles with low pressure grout pumps at a pressure not exceeding the designed strength of the concrete. The grouting shall be started at very low pressure and increased gradually to a required pressure. The grouting shall continue till the hole refuses to take any further grout, even at an increased pressure.

- The water soluble monomer based chemical used shall conform generally to IS:2645-1975*, specification for integral cement waterproofing compound.
- The nozzles shall be removed 24 hours after the grouting is over and the holes shall be finished off neatly.
- Inside surface shall be plastered with a layer of 1:3 cement mortal (1 cement: 3 Coarse sand) with waterproofing compound finished neat and properly cured.
- Work should be preferably executed by specialised agency that shall stand a warranty for 15 years for work executed.

(v) Guniting: In the case of leakages from very many points, guniting has to be resorted to for covering the entire surface with sufficient thickness of mortar strengthened with mesh reinforcement.

11.4 TROUBLESHOOTING

(a) Injection grouting to stop the entry of water

(b) Seal the leaking joints using sealants

(c) Apply waterproofing chemicals/membranes where the old one has given way or is non-existent.

(d) Provide Damp proof course to stop rising of water through capillary action. New DPC layer can be inserted in existing buildings.

(e) Improve drainage of surrounding areas if found to be inadequate.

11.5 NEW MATERIALS FOR PREVENTION/ RECTIFICATION OF LEAKAGE/SEEPAGE AND WATERPROOFING

It is important to choose right products and system to achieve a durable and impermeable concrete structure which can last more than its anticipated life. A number of chemicals, sealants and membranes/

treatments are now available for application of the wet areas and stopping water ingress.

Listed in Annexure B are some of the compounds/chemicals, which may be used as found suitable for the structure or site conditions keeping in view the manufacturer's recommendations. Other than these compounds/chemicals more materials are available in the market which may also be used as found suitable and economical for the structure/site condition as per manufacturer's recommendations.

BIBLIOGRAPHY

1. Maintenance and Quality Control in Civil Works: CPWD Training Institute.
2. Civil Engineering Review Magazine.
3. Khanna, P.N. Indian Practical Civil Engineers Handbook:
4. Deolalikar S.G. Plumbing Design and Practice.
5. Handbook on Water Supply and Drainage with Special Emphasis on Plumbing: Bureau of Indian Standards.
6. CPWD Specifications, 2009.
7. Panchdhari A.C. Inspection and Quality Control in Building Works.
8. Waterproofing and Damp-Proofing of Wet Areas in Building: Recommendations—IS 13182: 1991.
9. Manual of Sewerage and Sewage Treatment: CPHEEO, Ministry of Urban Development.
10. Manual on Water Supply and Treatment: CPHEEO, Ministry of Urban Development.
11. Panchdhari A.C. Water Supply and Sanitary Installations within Building Design, Construction and Maintenance.
12. Uniform Plumbing Code.
13. Plumbing Engineering Services Design Guide of "The Institute of Plumbing".
14. National Building Code of India 2005.

ANNEXURES

ANNEXURE A:
List of Important Indian Standards for Reference

651 - 1992	:	Specification for Salt glazed Stoneware Pipes and Fittings.
774 - 1984	:	Specification for Flushing Cisterns for Water Closets and Urinals (Valveless Siphonic Type).
775 - 1970	:	Specification for Cast Iron Brackets and Supports for Wash Basins and Sinks.
782 - 1979	:	Specification for Caulking Lead.
909 - 1992	:	Specification for Underground Valve Type.
1536 - 1989	:	Specification for centrifugally Cast (Spun) Iron Pressure Pipes for Water, Gas and Sewage.
1537 - 1976	:	Specification for Vertically cast Iron Pressure Pipes for Water Gas and Sewage.
1538 - 1976	:	Specification for cast iron fittings for (pt. 1 to 23) Pressure Pipes for Water Gas and Sewage.
1703 - 1989	:	Specification for Ball Valves (Horizontal Plunger Type).
1726 - 1991	:	Specification for cast Iron Manhole Covers and Frames-Part II :Specific Requirements for H.D. Circular type.

1729-1979	:	Specification for Sand Cast Iron Spigot and Socket soil Waste and Ventilating Pipes Fittings and Accessories.
1742 - 1983	:	Code of Practice for Building Drainage.
1795 - 1982	:	Specification for Pillar Taps for Water Supply purposes.
2326 - 1987	:	Specification for Automatic Flushing Cisterns for Urinals.
2548 - 1983	:	Specification for Plastic Water Closet Seat and Covers.
2556 (Part VI/		Part VI: Specific Requirements of Urinals Section 2 Half
Sec.2) - 1974		Stall Urinals.
2556 (P XI) - 1979	:	Specification for Vitreous Sanitary Appliances (Vitreous China) - Part XI Specific Requirements for Shower Rose).
2556 (Part-XIV) -	:	Part XIV: Specific Requirements of integrated squatting
1974		Pan.
3989 - 1984	:	Specification for Centrifugally Cast (Spun) Iron Spigot and Socket Soil, Waste and Ventilating Pipes Fittings and Accessories.
4984 - 1987	:	Specification for High density Polyethylene Pipes for Potable water supplies.
4985 - 1988	:	Specification for Unplasticised PVC pipes for potable water supplies.
5913-1989	:	Method of test for asbestos cement products.
7231 - 1984	:	Specification for Plastic Flushing Cisterns (Valveless Siphonic type) for Water Closets and Urinals.

13592 - 1992	:	Specifications for SWR pipes
13983-1994	:	Stainless steel sinks for domestic purposes specifications.
14735 - 1999	:	Specifications for SWR moulded fittings.

ANNEXURE B:
New Materials for Prevention/Rectification of Leakage/Seepage and Waterproofing

Product and Type	Application Area	Remarks
Manufacturer 1. AMES Research Laboratories Inc; Ames Research Laboratories, Inc., PO Box 1350 Jefferson, Oregon 97352 USA Toll-Free: 1-888-345-0809 PST, Phone: 503-588-3330 PST, Fax: 503-364-2380, Email: amesstaff@amesresearch.com		
Product and Type	*Application Area*	*Remarks*
Ames Pure Acrylic Plastic Primer	For priming the roofs and areas after surface preparation for receiving the next treatment.	Application on roofs. Costs with all applications and five gallons per 100 sq ft application at $ 1.40 per sq ft.
Seam tape	It is peel and seal rubber tape to seem all cracks.	
Super Elasto Barrier	Maximum stretch elastoeric acrylic plastic in liquid form. Applied 2 - 5 gallons per 100 sq ft.	
Manufacturer 2. KRYTON Group of companies, Canada, Kryton Buildmat Co, Pvt. Ltd India Contact Mr. Akhil Kakkar; Tel: +91 124- 438-1140 / 41; Fax: +91 124 -438-1142; Sales: sales@krytonbuildmat.com; Email: info@krytonbuildmat.com www.kryton.com; Website: www.krytonbuildmat.com		
Product and Type	*Application Area*	*Remarks*
Krystol T1 and T2; cementitious brush applied treatment	Crystalline technology, can be used on new and old surface for waterproofing of concrete from positive side and negative side as well.	Can be used on negative side too. Spray painting possible. Construction joints can also be treated.

(Contd.)

Water stop system	Used at construction joints	Concrete must be saturated before application.
KIM 9Krystol Internal Membrane	As admixture in concrete for waterproofing 2% by weight of cement	
Manufacturer 3. WEBEC Corporation; Tel: 1-877-509-3239 or +1-408-435-7048 for more information., www.webac.com; Local distributors are Mr. Nayan Bhuva, Cemseal System and Sales, Shri Sai Sidhi, Liberty Garden Rd No 1; Malad (West), Mumbai - 400064 Tel: 022 32507180, 0934217078		
Product and Type	*Application Area*	*Remarks*
Flex Coat flexible membrane	For roofs.	
Flex cast	For crack filling and in expansion joints.	
WEBAC®5621 1-C High-Build Bituminous Coating Surface Sealing	Mainly used for sealing basements underground car parks, damp rooms, supporting walls, terraces and balconies prior to applying the finish.	
WEBAC®5611 Bituminous Acrylate Dispersion Surface Sealing	WEBAC®5611 seals surfaces reliably against water pressure of up to 7 bar when applied at only 1 mm film thickness.	
WEBEC 151 and 157 are polyurethane water activated foams	For stopping water instantly. Used in concrete cracks and injections behind structures to create water proof membrane.	It is hydrophobic grout using very little water

(*Contd.*)

WEBEC 1403 PU -two component resin	For hair line crack injections	
WEBEC FLEXGEL	Manhole/sewer pipe injections	Hydrophilic grout requiring large qty of water to form foam /gel
Manufacturer 4. CICO Technologies Limited, S-38 Okhla Industrial Area Phase-II, New Delhi-110 020, Tel.No.91-11-4050 9400, Fax No.91-11-4050 9413; E-mail:cicotech@cicog roup.com; Website: www.cicogroup.com		
Product and Type	*Application Area*	*Remarks*
Tapecrete P 151, acrylic based Polymer Modified Cementitious composite coating system.	Waterproofing of wet areas, bathrooms, kitchens, etc. Can also be used for roofs with glass fibre reinforcement.	0.375 kg of polymer and 0.750 kg of cement used per sq m of concrete surface in two coats.
Manufacturer 5. KEMPER SYSTEM (INDIA) PVT. LTD., Indo-German Technology Park, S. No. 297, 298, 299, At Post Village - Urawade Tal. Mulshi, Dist. Pune 412 108, Tel.: +91-020-6674 0649 / 50 / 51 / 53, Fax: -6674 0652, E-mail: info@kemperindia.com, Web: www.kemperindia.com		
Product and Type	*Application Area*	*Remarks*
KEMPEROL AQUA	To seal sloping roofs and balconies as a waterproofing membrane underneath the tiles. Also for sealing of pipes when taken through cores	Not Alkali restatnt. To be used over primer.

(*Contd.*)

Product and Type	*Application Area*	*Remarks*
KEMPEROL V 210 one component ready to use membrane with long-term elasticity	As a waterproofing system for flat roofs, plant holders, buildings, etc.	Approx. 0.5–1.50 kg/m^2
KEMPEROL 165 Fleece 3-component water-proofing system based on flexible polyester resin.	As reinforcement for KEMPEROL waterproofing systems.	Approx 2.8–3.6 kg/m^2 giving a coating thickness of approx. 2 mm
KEMPERTEC EP- Primer, Special fleece based on polyester.	House hold Bathrooms, Shower installations and Wash rooms. Used with V 210	Approx 165 g/m^2
KEMPEROL 022, Solvent-free, liquid synthetic material		Approx 100 to 150 ml/m^2
Kemper system costs Rs. 180 - 200/- per sq ft and hence to be used in critical areas. Corners of roof, pipe joints, etc., where other applications may not give good results. On roof the taken up and coated with membranes of other treatments can be Keper treatment only in corner as an overlap to seal the area.		

(Contd.)

Manufacturer 6. STP Limited, 3rd Floor, Farm Bhawan, 14-15 Nehru Place, New Delhi - 110019; Tel: +91 11 46561359; Fax: +91 11 46561358; E-mail: co@stpltd.com		
Product and Type	*Application Area*	*Remarks*
Super Thermlay, APP Membrane SuperTene I, membrane	For roofing and basements	With Non woven Polyester Rft
Shalimar Supercrete	Waterproofing of wet areas, bathrooms,	Self sticking
Shalicrete 42	Bonding and waterproofing agent for repair.	Acrylic based polymer for multipurpose use
	Bonding and waterproofing agent for repair.	Acrylic bonding/Waterproofing compound
Manufacturer 7. Don Construction Chemicals India Ltd. (H O), No.4, II Block, Chowdhary Complex, 5th Street, Nandanam Extension, Chennai - 600 035. Phone: +91-44-2433 1817, +91-44-2433 5345, +91-44-5211 0993; Fax: + 91- 44- 2433 8272; E-Mail: dccil@vsnl.com; marketing @donconstruction.co.insilicon water		
Product and Type	*Application Area*	*Remarks*
Set Seal - Single Component Cementitious Coating,		Manufacturers of Admixtures, Waterproofing systems, Grouts, Adhesives and repair materials.
Set plug rapid set plugging mortar	For instant stopping of gushing water.	Fast setting hydraulic compound.
Repicone S -silicon water repellent	Barrier to water penetration on porous surfaces.	4–5 sq m per litre coverage.

(*Contd.*)

Manufacturer 8. BASF Construction Chemicals (India) Pvt. Ltd, C-68 MIDC Thane Belapur Road Turbhe Navi Mumbai - 400613; Tel: +91 22 4157 7777; Fax: +91 22 4157 7766; E-mail: basfcc@vsnl.net

Product and Type	*Application Area*	*Remarks*
Products include Performed membrane, Liquid Applied Membrane, Performed sealing Products, Liquid applied Sealing products	For sealing of joints, Waterproofing of roofs, basements and water tanks, etc.	Coniroof is without laps or welding
Master seal 505 Rapid Hardening Leak Plugging Mortar	Construction Joints, repair of cracks	Sets in 2 minutes
Master seal 550Polymer modified cementitious waterproofing coating	Sunken areas, Balconies and protection of concrete	Good bond strength
Master seal 550 EL Flexible Polymer modified cementitious waterproofing coating	As coating on concrete of basements, terraces and protection of concrete.	Has good elongation
PVC/TPO Membrane	For basements	
Coniroof liquid applied membrane	For roofs/ terraces	

(Contd.)

Manufacturer 9. Industrial Plants and Waste Treatment Corporation (IPWT) (New Delhi Branch Address 25, National Media Centre, NH-8, Gurgaon-122 022, Phone: +91-0124-400 8233; Fax: +91-0124-235 6906		
Product and Type	*Application Area*	*Remarks*
TAMMS "HEY'DI" Flexible Matrix and cementitious products	In depth Crystallisation Waterproofing system.	Flexible version of HEY'DI available to allow covering of cracks up to 1 mm. No protection necessary before back filling
ZUPA Derma Rubber, PLASTIFECT - PU, liquid applied elastomeric waterproofing membranes	For terraces	For locations where temperature variations are high
ZUPA, AQUATHON,	Waterproofing of concrete and masonry surfaces.	
Manufacturer 10. Colossus Construction Chemicals, Cemseal System and Sales, Shri Sai Sidhi, Liberty Garden Rd No 1; Malad (West), Mumbai - 400064 Tel: 022 32507180, 0934217078		
Product and Type	*Application Area*	*Remarks*
Polflex WP Polymeric flexible waterproofing coating	Waterproofing of Water Tanks and Toilets - sunken areas	Cost Rs. 45 - 50 per sq ft.
Conproof IWC	Waterproofing compound	

(*Contd.*)

Conpbond Acrl SP	Co - polymer acrylic for waterproofing.	Rs. 110/- per litre covers 30 sq ft in two coats.
Colossus sealing tapes and Floor packings	Acrylic based tape for expansion joint sealings below ceramic. Floor packings for sealing of drains below ceramic	
Manufacturer 11. Pidilite Industries Limited, 6th Floor, Vikas Deep, Laxmi Nagar, District Centre, Vikas Marg, DELHI - 110092; Phone :(011) 45529000; Fax: (011) 22422916, 22500720		
Product and Type	*Application Area*	*Remarks*
Dr. Fixit Pidfin 2K (water proof coating).	For coating of internal and external plastered surfaces.	It is polymer modified cemntitious seamless waterproofing membrane.
Pidiproof LW (Integral Liquid Waterproofing Compound)	Used as integral liquid waterproofing admixture.	
Dr. Fixit TORCHSHIELD roll	Applied on roof and basements over primer with torching.	Overlaps and sealing and protection in case of accebile roofs.
Dr. Fixit PIDICRETE MPB (Acrylic based bonding agent)	1. As a bonding aid for old and new concrete / mortar. 2. Useful as an additive for making repair mortars / concrete 3. water resistant render or plaster.	
Dr. Fixit PROOFCOTE BLACK (Heavy duty bituminous elastomeric waterproofing coating	It can be applied in sunken portion of toilets, sewage pipes, plinth, foundation and basement.	

(*Contd.*)

Dr. Fixit INSTANT LEAK PLUG (Fast setting powder for drip leakage)	Basement, concrete pipe, water retaining structure and RCC water tank.	Setting time is 3 min.
Manufacturer 12. Pidilite Industries Limited, Head Office- Ramkrishna Mandir Road, Off Sir Mathuradas Vasanji Road Andheri (East), Mumbai-400059 Maharashtra. INDIA; Phone: +91 22 3308 7000; Fax: +91 22 2835 7700; E-mail: pil@pidilite.com		
Product and Type	*Application Area*	*Remarks*
ROFF Hyseal (Surface applied waterproofing compound)	Concrete water tanks, sunken slab of bath room and toilets, basement, underground structures.	The chemical penetrates into water bearing capillaries, reacts with cement particle and form insoluble crystals.
ROFF Hytite (Water-proofing powder for concrete and mortar)	A chloride free powder that seal the surface of concrete and cement mortar there by blocking water seepage.	500 gm per bag of cement.
ROFF Hygrout (Acrylic Reinforced Cementitious, flexible waterproofing coating).	Injection grouting for the structure by filling voids.	Mix with water in the proportion 1:1 and pressure grout it into concrete.
ROFF Hyinject (Chemical based injectable damp proofing compound).	For external and internal wall that has gathered dampness due to rising ground water through brick masonry.	For injecting - 0.3–0.7 metre / litre, For wall coating 0.5–0.6 metre / litre (2 coat)

(*Contd.*)

Product and Type	*Application Area*	*Remarks*
ROFF Hydamp (colour enamel coating for water-proofing and protective treatment).	1. Ideally suited for application on plaster, bricks and blocks. 2. Waterproofing of parapets, outer walls, chajja and internal and external plastered surface.	
ROFF Hylast (colourless surface applied resin for waterproofing concrete surfaces, brickwall and asbestos, etc.)	For external vertical surfaces, brickwork, cement rendering, concrete blocks, weathering asbestos sheets and certain types of natural and artificial stones.	It should not be used below DPC level or below ground level.
ROFF Hyguard (Acrylic Reinforced Cementitious, flexible waterproofing coating).	1. As a waterproof lining of water retaining structure. 2. Sealing and coating tie bars holes to ensure water tightness. 3. Provide foundation protection and waterproof coating for roof. 4. Backing to marble and	
granite to prevent water ingress and thus alleviate surface staining.		
ROFF Hypel (Silicon based water repellent coating / rain coat.)	It is applied on siliceous/ mineral surface especially on exterior building brickwork, concrete cement / sand rendering, plastering, natural or cast stone.	

(Contd.)

Product and Type	*Application Area*	*Remarks*
ROFF Hyproof (water reducing, liquid integral waterproofing admixture for concrete.	It contains active ingredients to make concrete waterproof that can even resists water through hydrostatic pressure or capillary absorption.	It increases durability, cohesive property, workability and surface finish of concrete.
ROFF Hyfex (Ready to use Polymer modified flexible waterproof membrane coating).	Over cracks when its width lies between 0.5 mm and 1.5 mm as a band aid treatment, flat or sloped concrete roof, basement, metal or asbestos sheet joints.	
ROFF Hyguard EX (Polymer modified Cementitious waterproof coating).	1. Breathable waterproof coating. 2. Bridges hairline cracks in concrete structure. 3. Waterproofing for water tanks, swimming pool, etc. 4. Exterior waterproofing for basement wall. 5. Terrace and terrace garden waterproofing. 6. Sunken portion of bathroom.	
Manufacturer 13. Fosroc Chemicals (India) Pvt. Ltd, "PSR ID" No. 38, III Floor 12th Cross, CBI Road Ganganagar North Bangalore-560032 India; Tel: +91 80 2355 1500/01/02/03; Fax: +91 80 2355 1510; Email: enquiryindia@fosroc.com Delhi Office-109, Vikrant Tower 1st Floor, Rajendra Place, New Delhi, Delhi 110008 Tel.: 011 45062000		
Product and Type	*Application Area*	*Remarks*
Proofex Engage (Pre applied waterproofing membrane which mechanically bonds to poured concrete)	Basement and substrates	Can be trafficed immediately after application

(*Contd.*)

Product and Type	*Application Area*	*Remarks*
Hydroproof (single component acrylic polymer for cement based waterproof composite coating membrane).	Below sloping screed or waterproof plaster on - Roof, terraces, End wall subject to rain and balconies.	
Brushbond Cool coat (polymer based solar reflective and insulating coat).	External and internal walls, roof slab and cold storage facilities.	
Brushbond (Acrylic polymer modified elastomeric water-proofing membrane coating for concrete and masonry surface).	Water tank,reservoirs, roofs and swimming pools.	
Brushbond RFX (High performance elastomeric cementitious waterproof coating)	It can be used on concrete, brick and block work substrates and is equally suitable for new and existing structure.	Used for waterproofing and to protect atmospherically exposed reinforced concrete structure from attack by acid gases, chloride ions, water and oxygen.
Brushbond TGP (Crystalline capillary waterproofing system for cementitous substrates).	1. Sewage treatment and water treatment plants, tank foundation, tunnel and manholes. 2. Terrace garden, balconies and sunshade. 3. Water holding structure.	1. It should not be applied when the temperature is 5 degree or falling. 2. Its full activation and effectiveness may require 2-3 weeks.

(Contd.)

Product and Type	*Application Area*	*Remarks*
Proofex Torchseal 3P and 4P (Torch applied water-proofing membrane).	It provides the waterproofing and damp proof membrane for concrete roof and floor slab, basement tanking, car park deck slab, concrete retaining structure, tunnels and subway.	
Nitoproof 600 (Liquid applied single component elastomeric water-proofing membrane).	It ideal for wide range of water/vapour proofing application such as foundations, basements, ground floor, suspended floors, car deck, roof terraces, balconies, water retaining structure and sewage works.	
Proofex SM (Polymer modified spray applied waterproofing membrane).	It can be used for basements, underground structure, roofing, terraces, balconies	
Nitoproof 100 /200 (Bituminuos waterproof protective coating)	Horizontal and vertical damp proof membranes in sandwich construction for floor, walls and roof. Internal lining of water storage tank.	

ANNEXURE C: Checklists

To ensure that the works of installations of piping, fittings and fixtures of water supply and sanitary installations have been done correctly and are in good shape a set of check lists have been prepared and are given here. These will be useful to staff working at field level.

Check Lists for Quality of Works: C-1 "Water Pipes, Soil Pipes, Rain Water Pipes, Storm Water Pipes"

Sl. No.	*Description of Activity*
1	Check the wall and floors of toilets has been plastered with waterproofing compound
2	Check the shaft wall has been cleared from all loose materials
3	Check for the painting of C.I. pipes with anti-bitumastic paint
4	Check the pipes for the approved make
5	Check for quality and size of the pipes as per spec., particularly weight
6	Check for quality of supports, fixtures, sealants, bolts, nuts, thread pads, collars, etc.
7	Check routing to coordinate with other services, if any
8	Check for compatibility of fasteners, sealant and sleeves
9	Check for waterproofing of joints
10	Check for painting of G.I. pipes with anti-bitumastic paint, rolling in sand and covering with polythene
11	Check for all joints for air tightness by pressure testing and smoke test
12	Check all terminal connection in relation to flow diagram

13	Check for wall tippas before starting water lines in toilets and kitchen
14	Check the position of diverters, pillar cocks, bib cocks, consist stop cock and angular stop cocks position with respect to tiles
15	Check the level of the rain water pipes in balconies with tiles
16	Check the level of the rain water pipes in slope with the terrace
17	Check for all operational valves, meters, etc. for effective controlling and operation
18	Check all the pipes are placed before closing it from inside toilet and kitchen
19	Check the ducts are cleaned and pipes are painted before closing brickwork
20	Check the area is cleaned for the succeeding contractor
21	Check for cutting of chase of pipes mechanically and not by chisel and hammer
22	Check for closing of G.I. pipe chases before plastering/fixing tiles
23	Check for proper covering of S.W. pipes with concrete
24	Check the provisions of cleanouts and levels of chambers/ manholes.
25	Check alignment and levels of the soil and drainage lines.

Checklists for Quality of Works: C-2 "Injection Grouting"

Sl. No.	*Description of Activity*
1	Check the fixing of grouting pipes in slab/wall. One pipe should not cover more than 0.33 cum of concrete.
2	In case of horizontal slab, the top of pipe should be covered.
3	In case of wall, check that grouting are fixed at the junction of lifts of casting.
4	In case of wall, check that all cracks holes, etc. in the face opposite to where grout pipes are fixed are properly sealed.
5	Check the consistency of the grout.
6	Check the grout mix that it is as per specifications and desired quantity of waterproof compound @ 1.0 kg or Vz kg per bag (as per specifications) is mixed.
7	Check the grout pressure (2 kg/cm^2).
8	Check the grout is not oozing out from some hole/crack.
9	Check that pumping is done till grout is accepted by the concrete.
10	Check that no holes or chases are cut after injection grouting.
11	Check for curing.

Checklists for Quality of Works: C-3 "Water Supply and Sanitary Installations"

Sl. No.	*Description of Activity*
1	All sanitary fixtures are of approved make and comply with BIS specifications.
2	HCI and CI pipes checked for dimensions, weight, finish and coating of anticorrosive treatment in factory.
3	Conformity of GI pipes, fittings and fixtures to BIS and specifications done?
4	Sizes of brackets for wash basins, cisterns and sinks checked.
5	Whether traps provided in wash basins and sinks.
6	Quantity and quality of lead in pipe joints.
7	MS Clamps used to fix HCI pipes checked for size weight and fixing in concrete blocks.
8	Dimensions of trenches for laying pipes their bedding and filling.
9	Prior to execution planning of the services has been done and depressions of floors, if provided are as per drawings and site requirements?
10	Water Storage tanks - thickness of sheets, whether covers are mosquito proof.
11	Manhole covers, road gully gratings, etc.—check for weight, size, make and finish.
12	Testing of joints of water lines and sewers done?
13	Refilling of trenches of pipes - consolidation of earth.
14	Depth of channels and benching in manholes.
15	Record of waterproofing compound
16	Checking of locations of floor taps, pipe out lets and fixtures to see suitability with respect to tile size and aesthetics.

17 Check that fittings are firmly secured.

18 Check that joints are properly made and sealed.

19 Ensure that there are no paint or mortar droppings on fixtures and appliances.

20 Check that all appliances are functioning properly.

21 Check that covers of sensors are in place and they operate properly.

INDEX

Made in the USA
Monee, IL
05 November 2026

67807969R00105